AUTOMATED BIOMONITORING:
Living Sensors as Environmental Monitors

ELLIS HORWOOD BOOKS IN WATER AND WASTEWATER TECHNOLOGY

This collection of authoritative works reflects the awareness of the importance to the world of water, wastewater treatment, and closely related subjects. The titles are written and edited by experts from a wide range of countries closely concerned with research and development, monitoring and improving water quality and supplies, and treatment and disposal of wastewater.

Titles published in collaboration with the WATER RESEARCH CENTRE, UK

AQUALINE THESAURUS 2
Editor: G. BOTHAMLEY, Water Research Centre
SEWAGE SLUDGE STABILISATION AND DISINFECTION
Editor: A. M. BRUCE, Water Research Centre
STABILISATION, DISINFECTION AND ODOUR CONTROL IN SEWAGE SLUDGE TREATMENT
Editors: A. M. BRUCE and E. S. CONNOR, Water Research Centre
BULKING OF ACTIVATED SLUDGE
Editors: B. CHAMBERS and E. J. TOMLINSON, Water Research Centre
BIOLOGICAL FLUIDISED BED TREATMENT OF WATER AND WASTEWATER
Editors: P. F. COOPER, Water Research Centre, and B. ATKINSON, University of Manchester Institute of Science and Technology
ENVIRONMENTAL TOXICOLOGY: Organic Pollutants
J. K. FAWELL and S. HUNT, Water Research Centre
ENVIRONMENTAL PROTECTION: Standards, Compliance and Costs
Editor: T. J. LACK, Water Research Centre
WATER RESEARCH TOPICS
Editor: I. M. LAMONT, Water Research Centre
SMALL WATER POLLUTION CONTROL WORKS: Design and Practice
E. H. NICOLL, Scottish Development Department, Edinburgh
EFFECTS OF LAND USE ON FRESH WATERS
Editor: J. SOLBÉ, Water Research Centre
RIVER POLLUTION CONTROL
Editor: M. J. STIFF, Water Research Centre

Other books in Water and Wastewater

WATER POLLUTION BIOLOGY
P. ABEL, Sunderland Polytechnic
ALKALINITY–pH CHANGES WITH TEMPERATURE FOR WATERS IN INDUSTRIAL SYSTEMS
A. G. D. EMERSON, Independent Consulting Technologist
CHEMICAL PROCESSES IN WASTE WATER TREATMENT
W. J. EILBECK, University College of North Wales, and
G. MATTOCK, Waste Water Treatment and Resource Recovery Consultant
BIOLOGICAL TREATMENT OF SEWAGE BY THE ACTIVATED SLUDGE PROCESS
K. HÄNEL, Dresden
ANALYSIS OF SURFACE WATERS
H. HELLMANN, Federal German Institute of Hydrology, Koblenz
WATER-ABSTRACTION, STORAGE, TREATMENT, DISTRIBUTION
J. JEFFERY, General Manager, North Surrey Water Company
POLLUTION CONTROL AND CONSERVATION
Editor: M. KOVACS, Hungarian Academy of Sciences
NATURAL WATER SAMPLING
J. KRAJCA, Research Insitute of Geological Engineering, Brno, Czechoslovakia
GROUND RESOURCES AND YIELDS
H. KRIZ, Institute of Geography of the Czechoslovak Academy of Sciences, Czechoslovakia
PREVENTION OF CORROSION AND SCALING IN WATER SUPPLY SYSTEMS
L. LEGRAND, Former Head of the Water Dept. of City of Paris, and P. LEROY, Chemical Engineer of City of Paris, Chief of Corrosion Lab.
HANDBOOK OF WATER PURIFICATION Second Edition
W. LORCH, The Lorch Foundation
WATER MANAGEMENT IN INDUSTRY AND COMMERCE
P. MILLINGTON, Consulting Engineer, Staffordshire
BIOLOGY OF SEWAGE TREATMENT AND WATER POLLUTION CONTROL
K. MUDRACK, University of Hanover, FRG, and S. KUNST, Institute for Sanitary Engineering, Hanover, FRG
IMPROVING WATER QUALITY
M. RAPINAT, Compagnie Générale des Eaux, France
DRINKING WATER MATERIALS: Field Observations and Methods of Investigation
D. SCHOENEN and H. F. SCHÖLER, Hygiene-Institut der Universität, Bonn, FRG
HANDBOOK OF LIMNOLOGY
J. SCHWOERBEL, University of Freiburg, West Germany
BIOTECHNOLOGY OF WASTE TREATMENT AND EXPLOITATION
Editors: J. M. SIDWICK and R. S. HOLDOM, Watson Hawksley, Buckinghamshire
INTERACTION OF MICRO-ORGANISMS AND CHEMICAL POLLUTION OF WATER
D. TOTH and D. TOMASOVICOVA, Insitute of Experimental Biology & Ecology, Bratislava
BIOLOGICAL TREATMENT OF WASTEWATER, 2nd Edition
M. WINKLER, University of Surrey

ELLIS HORWOOD BOOKS IN AQUACULTURE AND FISHERIES SUPPORT

Series Editor: Dr. L. M. LAIRD, University of Aberdeen

BACTERIAL FISH PATHOGENS: Disease in Farmed and Wild Fish
B. AUSTIN and D. AUSTIN, Heriot-Watt University
MICROBIAL BIOTECHNOLOGY: Freshwater and Marine Environments
B. AUSTIN and D. AUSTIN, Heriot-Watt University
MICROBIOLOGICAL METHODS FOR FISH AND SHELLFISH
B. AUSTIN and D. AUSTIN, Heriot-Watt University
AQUACULTURE, Volumes I & II
G. BARNABÉ, Université des Sciences et Techniques du Languedoc
AUTOMATED BIOMONITORING: Living Sensors as Environmental Monitors
D. GRUBER and J. DIAMOND, Biological Monitoring Inc., Virginia, USA
SALMON AND TROUT FARMING
L. M. LAIRD and T. NEEDHAM, University of Aberdeen

AUTOMATED BIOMONITORING:
Living Sensors as Environmental Monitors

D. GRUBER, Ph.D., M.S., B.S.
President and Technical Director

and

J. DIAMOND, B.A., M.S., Ph.D.
Chief Scientist

both of Biological Monitoring Inc.
Blacksburg, Virginia, USA

ELLIS HORWOOD LIMITED
Publishers · Chichester

Halsted Press: a division of
JOHN WILEY & SONS
New York · Chichester · Brisbane · Toronto

First published in 1988 by
ELLIS HORWOOD LIMITED
Market Cross House, Cooper Street,
Chichester, West Sussex, PO19 1EB, England
The publisher's colophon is reproduced from James Gillison's drawing of the ancient Market Cross, Chichester.

Distributors:
Australia and New Zealand:
JACARANDA WILEY LIMITED
GPO Box 859, Brisbane, Queensland 4001, Australia
Canada:
JOHN WILEY & SONS CANADA LIMITED
22 Worcester Road, Rexdale, Ontario, Canada
Europe and Africa:
JOHN WILEY & SONS LIMITED
Baffins Lane, Chichester, West Sussex, England
North and South America and the rest of the world:
Halsted Press: a division of
JOHN WILEY & SONS
605 Third Avenue, New York, NY 10158, USA
South-East Asia
JOHN WILEY & SONS (SEA) PTE LIMITED
37 Jalan Pemimpin # 05–04
Block B, Union Industrial Building, Singapore 2057
Indian Subcontinent
WILEY EASTERN LIMITED
4835/24 Ansari Road
Daryaganj, New Delhi 110002, India

British Library Cataloguing in Publication Data
Gruber, D. (David), *1947–*
Automated biomonitoring: living sensors as environmental monitors. —
(Ellis Horwood books on water and wastewater technology).
1. Water. Pollution. Biological monitoring
I. Title II. Diamond, J. (Jerome), *1951–*
363.7′39463

Library of Congress Card No. 88–11204

ISBN 0–7458–0310–5 (Ellis Horwood Limited)
ISBN 0–470–21141–5 (Halsted Press)

Phototypeset in Times by Ellis Horwood Limited
Printed in Great Britain by Hartnolls, Bodmin

Table of contents

Preface

The contents of this book stem from a workshop on Automated Biological Monitoring Systems held during the annual Society of Environmental Toxicology and Chemistry (SETAC) meeting, November 4, 1986 in Alexandria, Virginia. Participants in that workshop, most of whom have contributed papers to this book, represented international and national interest. In fact, one of the most gratifying aspects of the workshop was the 'global' participation and the commonality of purpose despite the diverse backgrounds. The workshop also served to indicate the degree to which water quality is a problem of global concern, and the significant contributions automated biological monitoring systems can make in dealing with this problem.

As co-leaders and participants in this workshop, we were aware of several interesting phenomena relating to the development and use of automated biological systems. First, we noted the greater acceptance and use of these systems in Europe and countries other than the United States. This may not be a remarkable phenomenon in view of the greater proportion of scarcely populated land and unimpacted water sources in the US than in many other countries. We feel that this difference in perspective is a product of greater awareness for clean water in other countries, perhaps suggesting that clean water there may be a finite resource. Now the US is increasingly aware of the need to protect their freshwater resources and to safeguard high water quality standards. As a result, we have observed a renewed interest in the US in biological monitoring of surface and ground water and specifically, interest in automated biological water quality systems.

A second phenomenon we noted in the workshop, was the increased use of automated biological systems in monitoring drinking water as opposed to industrial effluents and wastewaters. This point is discussed elsewhere in this book but deserves mention here to indicate the trend to utilize these systems for the purpose of providing more immediate use and protection for mankind. This trend is occurring in all countries, including the US, and we see that this trend will probably increase further in the near future.

The third and last observation from the workshop worth noting is the fact that all of the participants (and others not present) are involved in a frontier science. All of the work presented at the workshop and in this book is an emerging form of biological technology which we feel will eventually serve as a prototype for the future benefit of mankind. As with any frontier science, developments are tested, refinements are made, and other existing technologies, first thought of as unrelated, suddenly become intimately related to the emerging systems. New forms of imaging and signal waveform technology, for example, which were not even envisioned 20 years ago, are inherent in some monitoring systems today. The work presented here is only the beginning of what promises to be a beneficial technology for mankind. For the first time, exciting possibilities are within reach.

Introduction

A storage tank, containing several hundred thousand gallons of fuel oil, accidentally ruptures, slowly spilling its contents into a nearby river. Many miles downriver and some days later, scores of dead fish are seen in the river and several health complaints are registered by people in nearby towns. It is discovered that the raw drinking water from this river is contaminated with fuel oil and therefore the water plants must close. Daily chemical monitoring will be required in order to determine when the water is safe to be used for drinking water again.

Twenty years ago, the above scenario may have seemed like wild fantasy, or at best a very rare and unfortunate incident. Today, however, far too many incidents like the one described have occurred throughout the world, sometimes with severe consequences. The catastrophic perturbations resulting from the chemical spill into the River Rhine in Europe or the spill of kepone into the James River and Chesapeake Bay in the USA are just two glaring examples showing the magnitude of the problem. The practice of continually monitoring surface waters, upon which man depends, is no longer a futuristic luxury but rather a necessity if we are to minimize potential catastrophes due to accidental (or planned) chemical spills.

As early as 1970, members of the scientific community, such as John Cairns Jr, were developing tools, using living organisms as sensors, whereby water quality could be measured continually and automatically. At the time these systems were first developed, the primary application was envisioned to be that of monitoring industrial wastewater discharges into surface waters. It was shown that some automated biological systems could provide continual information regarding the potential environmental hazard of a particular wastewater. From a management and regulatory point of view, this appeared to be an ideal system for monitoring the toxicity of a given wastestream since the system was always monitoring the effluent.

Shortly after the initial development of automated biological systems, several other workers around the globe began researching these systems and refining them to the point that they could adequately serve as biological early warning systems

(BEWS). That is, these biological systems could rapidly detect the presence of acutely toxic pollutants in a given wastestream almost as they occurred. As a result of this refinement, many of these automated biological systems could now be used not only to monitor water quality quickly and continuously, but also to provide enough warning such that remedial or corrective action could be taken. The remedial action might be diverting the wastewater to a holding pond for further treatment (to reduce toxicity), or closing the water intake to a drinking water plant until the hazard was shown to be removed.

A comparison between earlier research on the subject of automated biomonitoring and the present ones points up the changes which have occurred in BEWS development. The earlier discussions centered around the need for effluent biomonitoring and collection of biological data in general. The systems described in the earlier studies were clearly developmental and in many cases only laboratory tested. In contrast, most of the systems presented in this volume have been field tested and often incorporate sophisticated electronic or computer technology which was unavailable even 10 years ago. As a result, today's systems are more streamlined, more sensitive, and more reliable than the earlier prototypes. For these reasons, the present systems have much wider applications than earlier versions.

In recent years, we have discovered that the primary application of BEWS has shifted away from effluent monitoring and towards drinking water or surface water monitoring. Many of the systems presented in this book have been developed to monitor and protect drinking water supplies. Such systems have now been tested and are in use in England, Germany, the Netherlands, South Africa, and the USA. We see this application of biological early warning systems as being increasingly important to mankind. As world urbanization and industrialization increase over time, the potential for water quality hazards and human risk will increase, thus increasing the need for clean water supplies. BEWS can warn of the presence of a hazardous situation, minimizing the extent of catastrophe and ensuring clean water supplies.

The papers presented in this book are divided into three general sections, although they are all interrelated. The chapters in the first section deal with conceptual and historical issues of BEWS and the ever-present problem of system validation. These papers provide a critical summary of BEWS as they have existed and as they exist now, and of where they should be heading in the future. The second section presents papers on BEWS utilizing fish as the biological sensor. These systems are by far the most numerous at this time. The third section presents BEWS using biological sensors other than fish. We feel that this area is likely to become increasingly important as more alternative sensors are tested and utilized.

The information generated by the development of BEWS such as that described in this book, is not limited to water quality monitoring only. As public interest in BEWS increases and sophisticated technology becomes more accessible, we envision that BEWS development will stimulate research in a number of scientific realms. Animal behavior research, for example, is intimately related to the development of many BEWS currently in use. It is likely that theoretical and applied research in animal behavior will be enhanced by development in BEWS and vice versa.

The degree of electronic and computer equipment sophistication now utilized in many BEWS may offer technological challenges, spurring the development of new approaches to electronic signal capture and analysis. It is conceivable that BEWS development could trigger more varied and precise ways to monitor remotely physiological functions and behavior visually and electronically. Such technology could in turn lead to further refinements in BEWS, medical diagnostic technology, and other fields where remote biological sensing is critical.

Finally, it is highly probable that current developments in aquatic BEWS may be instrumental in the design of automated environmental monitoring systems for use in space stations, underwater outposts, and other man-made environments where air and water quality must be monitored. The body of work presented in this book represents a frontier in biological science which may well prove to be the foundation for a new technology beneficial to all mankind.

Part I

Perspectives and concepts in automated biological systems

1

A historical perspective

David Gruber,
Biological Monitoring, Inc., PO Box 184, Blacksburg, Virginia 24060, USA

The use of biological monitoring techniques for the purpose of protecting the waters of the United States is a most recent phenomenon. Consider, that it was not until the 1984 US EPA Policy Statement that the utilization of biological monitoring techniques for the purpose of monitoring potentially toxic wastewaters became so highly emphasized. Up to that point, biological monitoring techniques were officially used for such purposes only 'as appropriate'. The use of automated biological monitoring techniques for protecting water supplies presently remains a novelty and a pioneering endeavor. Nonetheless, just as the regulated community has accepted the use of biological techniques for toxicity assessments, the application of automated and real-time biomonitoring techniques, if kept cost effective and reliable, may become an inevitable reality.

To the scientific community, the term biological monitoring is a means of observing or keeping track of living organisms for a special purpose. With this definition in mind, historical interest in biological monitoring techniques undoubtedly preceded recordings in the literature. Even the Old Testament tells us that fish without fins or scales (e.g. shell fish) should not be eaten. Obviously, biological monitoring of these organisms led those people to recognize the feeding behavior of such organisms, and they concluded that these animals were not fit to eat. The term biological monitoring has been given more specific connotations by the scientific and regulatory communities as applied to issues of environmental and health concerns. For these applications, biological monitoring is a means of assessing water quality, or a chemical, for relative toxicity, utilizing living organisms as the sensors. Automated biological monitoring techniques are those which are self-acting or self-regulating. There is a fine line between biological monitoring studies conducted from the perspectives of basic science and applied science. In reality, the earliest studies were not concerned with applications for protecting the environment, but rather were concerned with the organism's response to a variety of stimuli.

1.1 FISH LOCOMOTOR ACTIVITY

Given the variety of developing automated biomonitors, this author feels best qualified to present a historical perspective of automated biomonitoring looking specifically at two of the present systems. The first system monitors the locomotor acitivity of aquatic organisms in an open field, as discussed in Chapters 8, 11 and 13. The earliest studies attempted to quantify the locomotor activity of aquatic organisms by visually observing their movements across a series of grids drawn on the bottom of the tank. Such techniques are often applied today in preference–avoidance studies. In this manner, certain behavioral variables, defined in terms of locomotor activity (e.g. time spent in a given area of the tank and number of entries) were quantified. In some cases, the investigators were able to quantify turning behavior as well, for example, the number of left- versus right-hand turns. However, such techniques required keen observations and the data often reflected subjective bias. With the advent of photographic and cinematic technologies, the subjective bias was reduced, but data handling remained tedious.

Automated remote monitoring of fish locomotor activity evolved because of the basic need for scientists to quantify significant changes in locomotor behavior. These could be grasped reliably only through long-term, uninterrupted observations yielding large volumes of data for meaningful statistical evaluations. Two unique systems for these purposes were developed by Dr Herman Kleerekoper, during the period of the early 1960s through the 1970s. These two systems appear to have had their published origins at McMaster University in Canada and subsequently evolved under Dr Kleerekoper's guidance at Texas A&M University. I personally had the good fortune of working with both systems while a PhD candidate with Dr Kleerekoper.

The first system was referred to as the 'circular tank' (Fig. 1.1). The circular tank had its published origins in 1961 (Kleerekoper *et al.*, 1961) and over the years it was developed through gradual improvements into five different tanks of two different dimensions (9 ft and 18 ft diameters). These cylindrical tanks were divided into sixteen compartments located along the perimeter of the tanks. Eventually, a light source and a bank of photocells were placed across the entrance of each compartment. Water flowed from the perimeters, through each compartment, and out through a standpipe located in the center of the tank. As a fish entered or left each compartment, the time and compartment number were recorded. Recording equipment evolved from photographic to computer-based sensing. Data were subsequently transferred from a paper-tape punch to a mainframe computer for analytical purposes. Locomotor activity was quantified into variables such as the number of entries into each compartment, time spent in each compartment, diurnal distribution of all movements regardless of direction, and the frequency with which any one of the 15 routes open to the animal on leaving a compartment was selected (Kleerekoper, 1969).

The second monitoring system was referred to as the square tank. This system resulted from Dr Kleerekoper's interest in monitoring the unrestricted locomotor patterns of fish in an open-field environment (Kleerekoper, 1967; Kleerekoper *et al.* 1970). Briefly, this system (Fig. 1.2) measured 5×5×p.5 m and possessed a matrix of nearly 2000 photocells embedded into the floor of the tank. An overhead light source

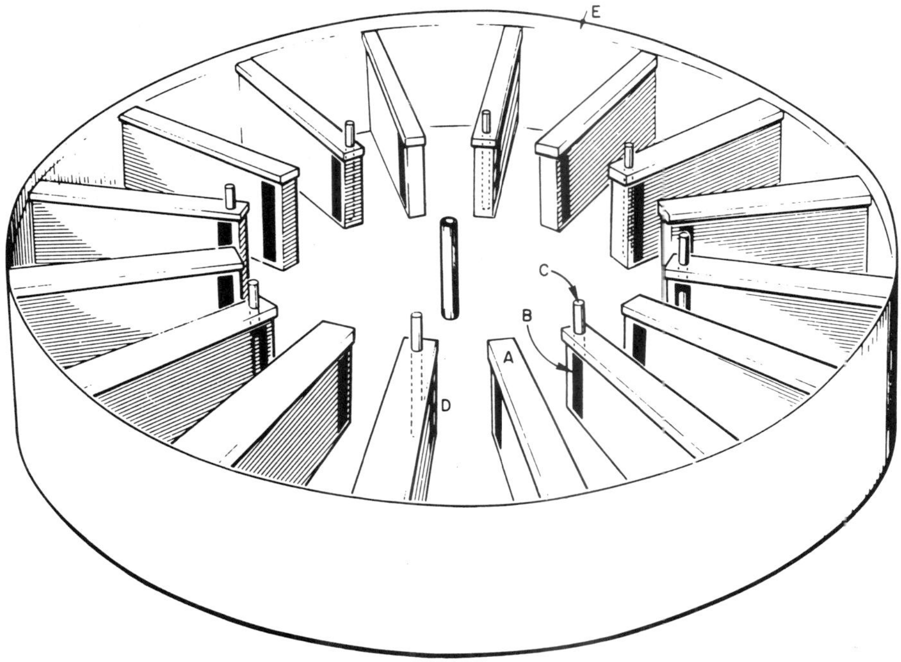

Fig. 1.1 — The 'circular' monitoring chamber employed for quantifying fish locomotor behavior (reproduced, with permission, from Kleerekoper, 1969).

created a shadow of the fish onto the photocells. A minicomputer monitored the fish's location (X–Y axes) and time. Data were subsequently transferred to a mainframe computer for data analyses. The locomotor variables which could be quantified included turning behavior (size, direction, frequency, sequence of angles of turns and the length of the steps which separate the turns), velocity, and the time spent in various areas of the tank.

Subsequently, these systems evolved from the use of photocells to the use of video cameras (e.g. Miller *et al.,* 1932). The digital output of video cameras facilitated their interface to computer-based systems. Contrast comparators in the interface enabled the identification of the location of aquatic organisms on an X–Y axes and time basis. The only limitation placed on the video-based systems is that the size of the open-field monitoring chamber must remain relatively small. With respect to the applications for monitoring for environmental health concerns, the small size of the monitoring chamber is advantageous. Consider, for example, the space limitations at water treatment plants. In Chapter 13, Smith and Bailey discuss more recent technological advances presently being used in locomotor systems.

1.2 FISH VENTILATORY BEHAVIOR

The second group of automated biomonitoring systems to be addressed historically are those which monitor the ventilatory behavior of fish. Belding (1929) appears to

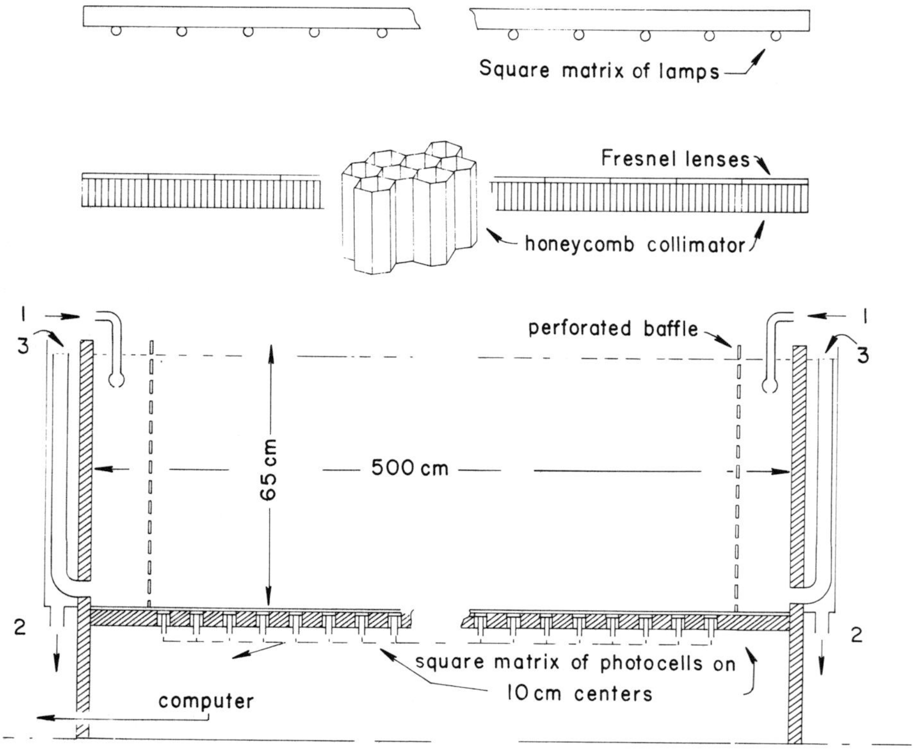

Fig. 1.2 — The 'square' tank. An open-field monitoring chamber utilized for quantifying fish locomotor behavior. (1) Water inlet; (2) water outlet; (3) standing pipe for water level regulation (reproduced, with permission, from Kleerekoper *et al.*, 1969).

have been cited most frequently as having begun the developments in this field with his studies on the respiratory movements of fish as an indicator of a toxic environment. Several investigators have played significant roles in the development of automation in such systems. Several chapters in this book discuss many of these developments and are written by some of these same investigators. However, recognition must be awarded Dr John Cairns Jr and his colleagues for contributing much impetus toward the concepts and developments of automated, real-time biomonitoring systems using fish ventilatory behavior for sensing toxic environments.

When Dr Cairns left the Academy of National Sciences and joined the staff of the University of Kansas, academic interests evolved into the development of a computer-interfaced system using fish respiratory rhythms and behavior to provide an early toxicity warning system for the industrial wastewater community. In 1968, Dr Cairns accepted a position as a University Distinguished Professor and his research continued at Virginia Tech University (Blacksburg, Virginia). The earliest on-site applications occurred during that time (Cairns *et al.*, 1970).

I came to Virginia Tech University as a Senior Research Associate in 1976 to study under the direction of Dr John Cairns Jr at the University's Center for Environmental Studies. Previously, I was completing my doctoral program requirements at Texas A&M University under Dr H. Kleerekoper. My activities with Dr Cairns quickly became centered around a unique area of academic and applied environmental interests. The US Department of Defense funded most of the research with which I was directly involved.

Perhaps the achievements of most significance to this book were the developments that led to the installation of a 'state-of-the-art' automated biomonitoring system at an army ammunition plant (Fig. 1.3). The system was more of a facility,

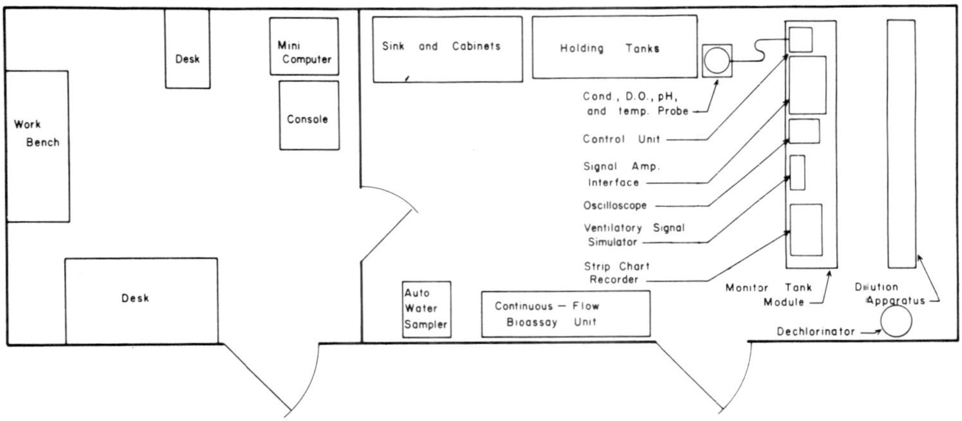

Fig. 1.3 — The internal layout of the monitoring and auxillary equipment within the biological monitoring facility which sensed fish ventilatory behavior as the monitoring parameter (reproduced, with permission, from Gruber *et al.*, 1979).

being housed within a sizeable trailer (32×12 ft). Considerations for industrial operators were incorporated into the design process of the facility. The system simultaneously monitored twelve fish, each individually housed within monitoring chambers and provided with automatic fish feeders and a dim light source. Two rows of 12 chambers were compartmentalized within a module (Fig. 1.4). The second row of monitoring chambers functioned to acclimate the fish to the apparatus. The use of the two rows of tanks alternately was designed to minimize the perceived potential for adaptation to exposure to low doses of toxins. The alarm routine could be triggered by the biological and/or physicochemical sensors. Upon generation of an alarm, buzzers would sound, warning lights would illuminate, a telephone call would be made and a message automatically generated, and, finally, a water sampler activated (Cairns and Gruber, 1979; Gruber *et al.*, 1981).

System validation became a major endeavor. Continuous-flow 96-hour acute bioassays were conducted during the periods used for experimental ventilatory monitoring. Use of exposure levels approaching 100% of that in the wastewater was not feasible owing to the acutely toxic nature of the effluent. Furthermore, the

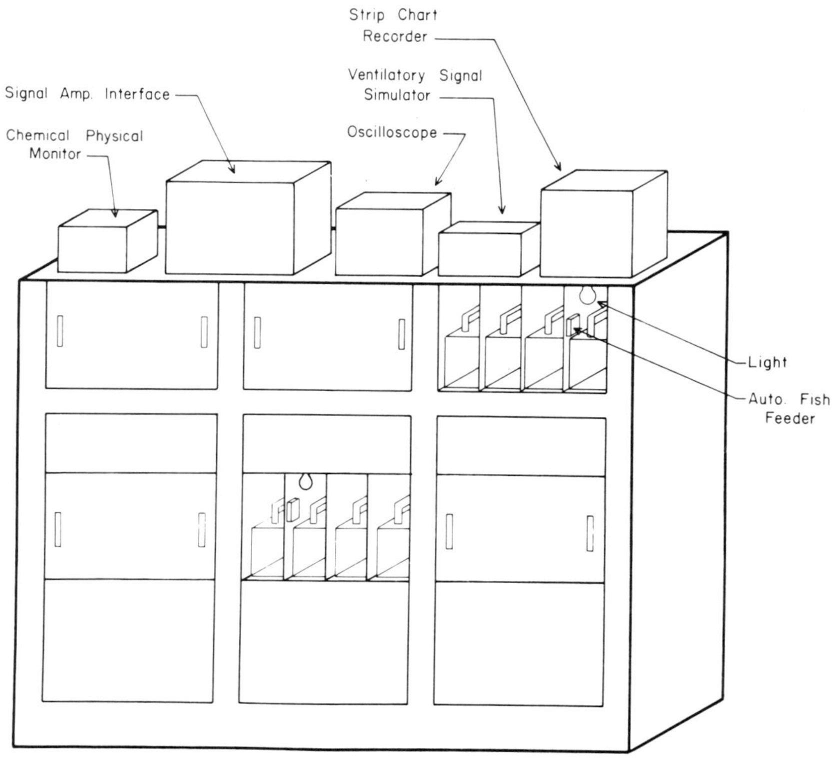

Fig. 1.4 — The compartmentalized module housing 24 monitoring chambers as utilized in the real-time automated biomonitoring system assessing fish ventilatory behavior.

variability of the effluent quality was extremely high. Exposure concentrations for the experimental runs were selected based upon the most recent LC_{50} (concentration lethal to 50% of test subjects) values available. At times, these dilutions were not appropriate, as was determined after the concurrent acute bioassays were completed. Nonetheless, subsequent analyses yielded a system capable of triggering alarm within 10 minutes of exposure to wastewater with toxicity levels approaching the LC_{50} values and within a day of exposure to levels of toxicity an order of magnitude lower than the LC_{50} value. We discuss this work in more detail in Chapter 2 of this book and elsewhere (Gruber *et al.*, in press). Other versions of automated systems using fish ventilatory behavior are discussed in Chapters 4–8 and 10.

1.3 RECENT DEVELOPMENTS AT BMI

In 1980 I had the unique opportunity of forming Biological Monitoring, Incorporated (BMI). BMI is a small business environmental consulting firm specializing in concepts of aquatic toxicology. It arose from the academic community in response to the needs of the regulated industrial and chemical communities. It is a classic example of the transfer of technology from the academic community to the applied environment. The continued research and developments in the field of automated biological monitoring were enhanced by the Federal Government's Small Business Innovative Research (SBIR) Grant Program.

This SBIR program was the result of the US Small Business Act of 1982, which was recently reauthorized. The result of this act was that each federal granting agency was required to set aside a small percentage (1–2%) of its annual grant program budget for funding innovative research proposals as received from the small business community. The program comprises three phases. The federal government funds phase I and phase II research provided that both the proposals and the results of phase I are received favorably. Phase III must be funded by non-federal monies with the eventual commercialization of a product. Phase I is typically a six-month 'think tank/feasibility' study, whereas phase II is the research and development phase of two year's duration. In this manner, the federal government funds the 'high-risk' early phases of such endeavors, thus facilitating the applications of innovative concepts fostered by the small business community. To date, BMI has received three such SBIR grants; a phase I from the National Institutes of Health, and a phase I and phase II grant from the National Science Foundation, which we gratefully acknowledge.

The latest developments in our automated systems are the result of research efforts conducted by BMI over the past six years. Briefly, the major accomplishments to date have included the scaling down of the monitoring chambers such that the use of juvenile fish is facilitated. The entire system requires a minimum of floor space, which also facilitates packaging and shipping. The use of the more recent computer technologies, such as the 286 generation computers, Turbo-Pascal-compiled software programs, and real-time analytical software capable of monitoring frequency, amplitude, and cough frequencies, has yielded a reliable and sensitive system. The alarm routines are packaged into the software so that both the biological and physicochemical parameters being monitored are graphically displayed in an appealing manner on a computer monitor with high-resolution graphics. The software is menu driven and extremely 'user friendly' with all queries defaulting to sensitivity levels determined during an initial calibration period conducted upon installation. Perhaps most appealing is that the cost of the system has been minimized to the point that it is equivalent to no more than the average cost for conducting a few on-site, 96-hour continuous-flow acute toxicity tests.

Our latest system is presently being installed at the municipal water treatment plant at the city of Los angeles, California. This unit is a stand-alone unit placed so that it is on public display. The City of Los Angeles receives its drinking water through an aqueduct system. The City's management have been acutely aware of their vulnerability to sabotage efforts and accidental hazardous spills which could contaminate their drinking water supplies. Their present chemical and physicochemical monitoring efforts are extensive and, for the most part, state-of-the-art. Yet, they are extremely cognizant of their responsibility to protect their public from contaminated drinking water. Their present monitoring technologies are unlikely to detect the presence of a water supply laced with cyanide or PCBs (polychlorinated biphenols) in time to prevent a major catastrophe. It is anticipated that BMI's automated biomonitoring system will act as one more means to minimize such risks. It seems that such applications point to the potential and inevitable acceptance and utilization of automated biomonitoring techniques for protecting waters in the US and abroad.

REFERENCES

Belding, D. L. (1929) The respiratory movements of fish as an indicator of a toxic environment. *Trans. Am. Fish. Soc.* **59** 238–246.

Cairns Jr, J., Dickson, K. L., Sparks, R. E. and Waller, W. T. (1970) A preliminary report on rapid biological information systems for water pollution control. *J. Water Pollut. Control Fed.* **42**(5) 685–703.

Cairns Jr, J. and Gruber, D. (1979) Coupling mini- and microcomputers to biological early warning systems. *BioScience* **29**(11) 665–669.

Gruber, D., Cairns Jr, J., Dickson, K. L., Hendricks, A. C. and Miller, W. R. III (1979). Initial testing of a recent biological monitoring concept. *Water Pollution Control Federation* **51**(11) 2744–2751.

Gruber, D., Cairns Jr, J. and Hendricks, A. C. (1981). Computerized biological monitoring for demonstrating wastewater discharge. *J. Water Pollut. Control Fed.* **53**(4) 505–511.

Gruber, D., Diamond, J. and Johnson, D. (in press) *Aquatic toxicology and hazard assessment: 11th symposium,* American Society for Testing and Materials. Philadelphia.

Kleerekoper, H. (1967) Some aspects of olfaction in fishes, with special reference to orientation. *Am. Zoologist.* **7** 385–395.

Kleerekoper, H. (1969) *Olfaction in fishes.* Indiana University Press, Bloomington.

Kleerekoper, H., Taylor, G. and Wilton, R. (1961) Diurnal periodicity in the activity of *Petromyzon marinus* and the effects of chemical stimulation. *Trans. Am. Fish. Soc.* **90** 73–78.

Kleerekoper, H., Timms, A. M., Westlake, G. F., Davy, F. B., Malar, T. and Anderson, V. M. (1969) Inertial guidance system in the orientation of the goldfish (*Carassius auratus*). *Nature* **223** 501–502.

Kleerekoper, H., Timms, A. M., Westlake, G. F., Davy, F. B., Malar, T., and Anderson, V. M. (1970) An analysis of locomotor behavior of goldfish (*Carassius auratus*). *Anim. Behav.* **18** 317–330.

Miller, D. C., Lang, W. H., Greaves, J. O. B. and Wilson, R. S. (1982) Investigations in aquatic behavioral toxicology using a computerized video quantification system. In R. B. Foster and W. E. Bishop, (eds) *Aquatic toxicology and hazard assessment: fifth conference,* STP 766. pp. 206–220. American Society for Testing and Materials. Philadelphia.

2

An overview of automated biomonitoring — past developments and future needs

J. Diamond, M. Collins, D. Gruber
Biological Monitoring, Inc., PO Box 184, Blacksburg, Virginia 24063, USA

2.1 INTRODUCTION

The search for a quick, reliable method of measuring environmental health of aquatic systems has been the focus of much research. Owing to this interest, short-term biological monitoring tools have been developed to assess the toxicity and water quality of wastewater discharges and receiving streams. These tests include acute and short-term chronic bioassays. A 48-hour test period is the minimum frame recommended in acute bioassays and seven days is the minimum duration for most short-term chronic bioassays (US EPA, 1985a, b).

Although much useful information has been generated using these short-term bioassays, these methods are not amenable to continuous real-time monitoring or useful for preventative monitoring purposes. Since information generated by laboratory toxicity tests may be available hours or days after test initiation, irreversible environmental damage could take place by the time the test results are known. In addition, if water or wastewater quality fluctuate over time, these tests may not adequately enable one to make reliable predictions concerning the risk of hazard to a receiving stream. Frequent water sampling and toxicity testing may be required in these cases, which could be very labor-intensive and costly.

In view of the above limitations, many researchers, regulators, and industries have observed the usefulness of developing a continuous, on-line biological early warning system, capable of rapidly detecting acutely toxic water conditions as they occur. In response to this need, several biological early warning systems (BEWS) have been developed over the past decade, many of which are described in subsequent chapters of this book. Although there is a wide diversity of BEWS currently under study today, they are alike in that all of them automatically monitor continuously expressed biological responses of aquatic organisms. Subtle changes in these responses over time provide an immediate indication of developing toxic conditions.

The fundamental components of a typical BEWS are shown in Fig. 2.1. The

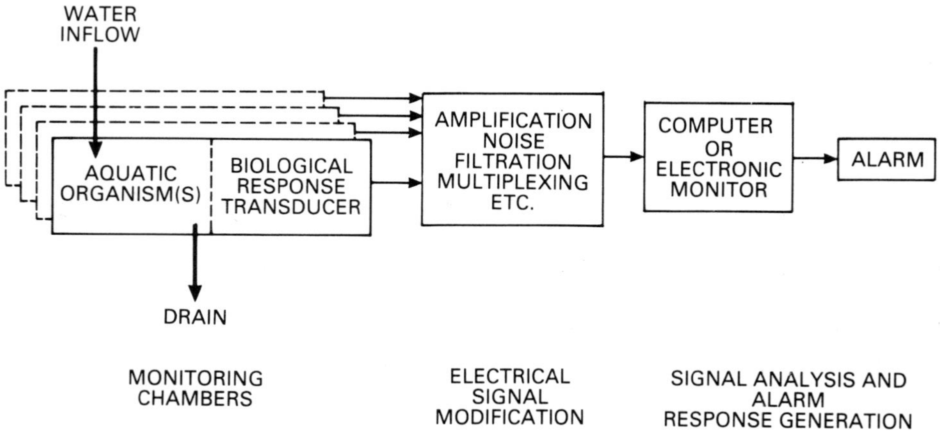

Gig. 2.1 — Fundamental components of an automated BEWS (after van der Schalie, 1985).

sensors (aquatic organisms) are exposed to continuous flows of the water being monitored. The biological response variable of interest is most often detected as an electrical signal or is converted to an electrical signal through the use of transducers. The biologically generated signals are sent to an electrical interface which converts the signals to an appropriate form for use by a computer or electronic monitor. Using some predetermined algorithm, the computer/monitor determines if the signals generated are indicative of toxic conditions. If toxic conditions are indicated, the system generates an alarm signal or initiates some other informative or corrective action (Cairns and van der Schalie, 1980).

Depending on the particular use of the early warning system (i.e. whether industrial wastewater, municipal wastewater discharges, or raw drinking water are being monitored), the specifics of the system and the criteria for evaluating the success of such a system can vary widely. Summarized below are specific criteria, initially developed for wastewater early warning systems, that apply equally well for systems designed to protect the quality of drinking water supplies and surface water. The following criteria include ideas and suggestions from Cairns and van der Schalie (1980).

(1) The operation of the system should be continuous and automatic. This requires that the biological response variable be quantifiable through computer analysis or other electronic recording equipment. It is desirable for this to be accomplished without having to restrain the test organism or directly attach devices to the organism. This avoids undue stress on the organism.
(2) The system must rapidly and reliably detect a wide-range of developing toxic

conditions. The response must be rapid enough for corrective action to be taken before adverse toxic effects take place. The greater the number of detectable toxic substances, the more valuable is the system.

(3) The number of false alarms due to non-toxic variations in water quality should be minimal. Certain characteristics of water (pH, hardness, dissolved oxygen) may change and elicit a response from an organism even though the change is non-harmful. Ideally, the system should be able to compensate automatically for such changes, thus avoiding false alarms.
(4) Appropriate methods for the analysis of data must be developed. The normal range of variation in the response variable should be statistically determined so that criteria for abnormal responses, caused by toxic conditions, may be established. Data analysis methods should allow for normal variation between individuals and diurnal variations.
(5) A monitoring system should be relatively easy to operate and the results should be easy to interpret. This is most easily achieved through automation via a computer system. Once such a system is in operation, most control functions could be automatic. This eliminates the need for highly trained personnel to be on-site continuously.
(6) The aquatic organism used as the sensor in the system should be fairly inexpensive and easily acquired and/or cultured at the facility. This limits the choice of organisms to those that are commercially generally available. This has the advantage that there is a large amount of literature on toxicity and culture methods for these organisms.
(7) The apparatus should require as little maintenance as possible. Complex mechanical arrangements should be kept to a minimum.

The above criteria represent what we consider to be ideals to pursue in developing a BEWS. In fact, however, most systems developed to date have encountered difficulties in attaining many of these ideals or in even addressing them. Historically, the emphasis for developing early warning systems stemmed from the desire to detect rapidly toxic chemical spills and sudden slugs of toxic substances arising from waste treatment or industrial accidents (Cairns *et al.*, 1970). For these purposes, emphasis on criteria (1) and (2) above (continuous, automatic, and rapid) were paramount to the design of the system. Since the goal of most of these systems was to detect acutely toxic conditions, a gross response in the sensor over time was usually obtained rapidly and definitively. Therefore, complex statistical algorithms and false-alarm routines were not often needed in the assessment process. As a result, in many of these systems, the issues of overall system sensitivity, reliability, and verification were not often addressed.

The importance of the above issues in the development of early warning systems is underscored by the relatively little attention thus far given to false alarm rate to non-toxic changes in water quality and appropriate statistical analyses to detect changes in sensor response. The deficiency becomes particularly crucial in situations where the early warning system is designed to detect sub-acute (chronic) levels in toxicity and where greater system sensitivity is needed. In these cases, the above criteria must be addressed in order to have a useful system.

We would argue that even in situations where the early warning system is designed to detect acutely toxic conditions, the issues of reliability and sensitivity must be examined in order to improve the usefulness of the system. This is because any system which generates a warning alarm must be able to account for the possibility of false alarms. In practice, this means that there must be some method to differentiate between alarms generated by toxic conditions and alarms generated by sudden changes in non-toxic parameters (pH, temperature, etc.). Therefore, the system would need to incorporate one of two general techniques: (a) a physicochemical system which automatically assesses whether the alarm generated is due to a non-toxic change (and therefore a false alarm) or (b) a statistical algorithm which accurately differentiates between changes due to non-toxic conditions and changes due to toxic substances. Although we recognize the need for physicochemical monitors in an early warning system, we prefer the second technique for determining false alarms since it is potentially more rapid and reliable then physicochemical data. Clearly this necessitates the development of sound statistical techniques and a sophisticated algorithm for assessing meaningful changes in the sensor response over time. Through an understanding of the basic biology of the sensor organisms and the type of responses likely to be elicited under varying conditions, such algorithms and statistical assessments can be defined.

Some basic conceptual issues which must be addressed more thoroughly in biological early warning system development are discussed below. These issues bear directly on the criteria outlined above but should not be taken to be the only issues of importance.

2.2 SPECIES SENSITIVITY

As outlined above, the sensor used in a given early warning system will depend on availability, cost, ease of maintenance, and sensitivity to toxicants and pollutants. Unfortunately, it appears unlikely that any one sensor species will satisfy all of these criteria for two reasons: (a) those species which are usually easy to culture and/or maintain under artificial conditions are often not the most sensitive to toxicants, and (b) no one species is equally sensitive to all possible pollutants. Fig. 2.2 illustrates the type of variation in sensitivity observed among typical indicator species to different toxicants. Species such as the fathead minnow and *Daphnia magna* are relatively sensitive to heavy metals such as cadmium and copper but relatively less sensitive to the pesticide dieldrin or benzene. Many workers have demonstrated similar differential sensitivities among species to a given toxicant and widely different responses to different types of toxicants using the same species.

The differential sensitivity of organisms to toxicants raises the question of whether a system employing a single species as the sensor is both reliable and sensitive enough to adequately monitor a particular surface-water body. Since a multiple species system is generally more complex in structure and operation then a single-species system, most workers up to now have concentrated on perfecting the single-species sensor alternative. One possible solution to the problem of species sensitivity using a single-species system is to incorporate a sensor which is known to be sensitive to the toxicants likely to occur in the water body being monitored. This of course presumes that the toxicants of concern are largely known beforehand. In

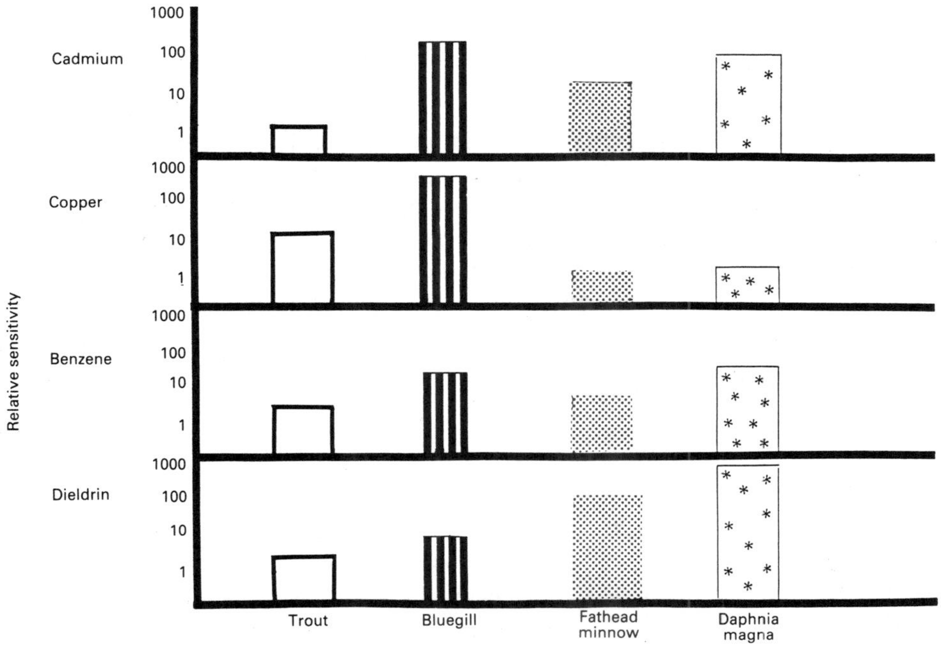

Fig. 2.2 — Relative sensitivity of various indicator species to certain pollutants.

situations where industrial wastewater discharges are being monitored, this may indeed be the case. However, for purposes of monitoring surface water where a variety of impacts could occur and the toxicants are not completely known, it may be impractical to designate one species as the optimal sensor.

The problem of differential species sensitivity to toxicants is not unique to biological early warning systems. In the realm of more traditional bioassays and water quality assessment, several ideas have been presented in dealing with this problem. For example, the US EPA assigns an uncertainty factor of 10 to account for differences in species sensitivity in order to assess whether further toxicity testing is necessary (US EPA, 1985c). The use of application factors in this regard is well established in hazard and risk assessment procedures (Dickson *et al.*, 1979). Owing to the lack of precision in the above approaches, several workers have cited the importance of devising multispecies systems in order to model better the environmental impact of a given wastestream or chemical (Cairns, 1983, 1984; Ewell *et al.*, 1986). Alternatively, other workers have indicated that certain indicator species may have in themselves reasonable predictive value for aquatic systems as a whole (Kenega, 1978; Kimerle *et al.*, 1983; Sloof *et al.*, 1986).

It would appear that the validity and reliability of using a single species as the sensor in an early warning system would depend on the proposed purpose of the system. If, for example, the system is intended to monitor surface water for protecting aquatic life, a single species, if chosen carefully, may be reasonably protective. Certain species, such as *Daphnia pulex*, could be overprotective sensors for a particular water body depending on the physical habitat and the biota indigenous to the system. Alternatively, if human health is being protected, a single species may or may not be protective, depending on the chemical constituents present in the water body and the species sensitivity to the potential toxicants. Chapter 3 discusses this issue, detailing the importance of validating the response of a BEWS in terms of the given water body being protected.

At the present, it appears that the problem of species sensitivity in early warning systems may need to be addressed on a case-by-case basis. One promising approach to this problem may be the development of a multisensor system which is relatively easy to maintain and operate, and in which the data can be easily analysed and interpreted. Prototypes of such systems by Dr Eric Morgan and coworkers (Morgan *et al.*, 1984; Chapter 9) and Dr W. S. Morgan (Chapters 7 and 10) are examples of what appears to be a fruitful line of research.

2.3 SENSOR ACCLIMATION

Sensor acclimation to toxicants over time is another conceptual issue of concern. It has been shown that organisms exposed to various toxicants at chronic levels for long periods of time can become acclimated to the toxicant (Miller *et al.*, 1986; Morgan and Kuhn, 1984; Cairns and van der Schalie, 1980). This can result in a lack of sensor response to otherwise harmful levels of a particular toxicant when they occur. The overall effect is reduced sensitivity of the system and decreased reliability.

It would appear that the problem of sensor acclimation can be readily solved by frequently rotating or replacing sensor organisms in the system. However, in practice, the problem may be less easily solved. For example, how will the user determine that the sensors are indeed acclimated to a level of toxicant and not simply unstressed? Furthermore, if the sensors do become acclimated over time to certain toxicants and appear unstressed, what then is the appropriate interpretation concerning the degree of impact in the water body? Finally, if sensors are rotated in the system to prevent acclimation, where will the sensors be maintained in the meantime? The above questions point up the importance of the sensitivity of the algorithm used to detect sensor response changes over time, and the careful planning needed in designing and running an early warning system.

2.4 HYDRAULICS AND TURN-OVER TIME

Factors affecting the time delay between the occurrence of toxic conditions and system response must also be considered. The location of the system is one factor affecting the time delay. The site should be located a minimal distance upstream (assuming surface water is being monitored) of the point of concern in order to include as many of the potential sources of contamination as possible. This could easily be accomplished in cases where an actual influent or effluent is being monitored. However, a second, and opposing factor to consider is that the location of an early warning system should be the maximum distance from the point of concern in order to allow sufficient time for data processing, establishment of alarm states, and implementation of subsequent emergency procedures. Clearly, the precise location of the early warning system will be a compromise between these two opposing factors, depending on the nature of the particular site. Other general considerations affecting the system location are that the site should be readily accessible to motor vehicles, acceptable to local authorities, and located above the flood plain. These factors may limit the optimal performance of the system as a whole.

Although proper placement of a biomonitor can give the unit more time to detect toxicity problems when they occur, there may still be a delay in response of the system. One reason for this is that the turnover time of the monitoring chambers must be considered. There may be a considerable lag before the toxic concentration in the chambers reaches that of the incoming water. This problem is particularly apparent in those systems utilizing larger sized sensors (large invertebrates and fish) where chamber size can be a factor. The lag time can be reduced by increasing the inflow rate and decreasing the chamber volume. The chamber volume, however, must not be decreased such that organisms are restrained with deleterious effects. Also, the increased flow could cause interference with some response signals. These problems would not be expected to be as relevant to those systems which use microinvertebrates or bacteria as the sensors.

Another potential source of delay between the actual occurrence of a toxicity problem and a system response is the fact that there exists an inverse relationship between biological response time and toxicant concentration. Any toxic event must be present long enough and at a great enough concentration to exceed the minimum detection level of the system. Thus, the system response delay must be considered on a site-specific basis. Considerations must include monitoring chamber turnover time, overall system response time, monitor intake location in relation to sources of possible contamination, and flow at the source intake.

2.5 SENSOR RESPONSE VARIABLE

Theoretically, any characteristic of an organism's normal behavior, which is evident continuously, could be used as a response variable in a biological monitoring system. However, in practice, there are several constraints concerning the choice of a suitable physiological or behavioral parameter for monitoring. The response monitored must be one that changes as a prelethal symptom of poisoning and it must be conducive to automation allowing for quick and reliable detection of deviations from normal conditions.

The appropriate response variable for an early warning system depends on the specific purpose for which the early warning system is designed. Fig. 2.3 presents schematically the sequence of organism responses with increasing toxicant concentration and exposure time. For example, given adequate technical means and expertise, changes in physiological and biochemical response variables, such as ventilatory profile, are likely to be detected under sub-lethal conditions more easily than observable phenomenon such as reduced growth rate or fecundity (Mehrle and Mayer, 1985; US EPA, 1985b). Likewise, changes in metabolic response variables such as oxygen consumption may be detected sooner, and under less toxic conditions, than more gross responses such as locomotory behavior or predator avoidance (Cairns and Dickson, 1980).

In order to have a truly rapid and sensitive system, it is evident that response

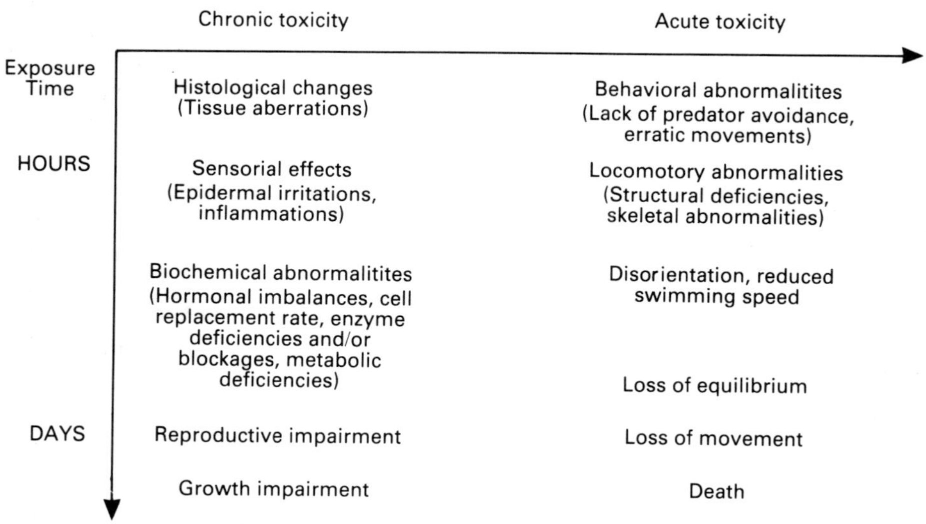

Fig. 2.3 — Theoretical scheme of organism responses to varying magnitudes to toxicity and over different time durations.

variables sensitive to a short exposure of toxic conditions are preferred over less sensitive response variables. How sensitive a response variable is, though, will depend in part on the available technology and its detection limits. Given the technology to date and the trends suggested in Fig. 2.3, this would mean that the use of physiological or biochemical response variables, for example, may be preferred over anatomical or behavioral response variables in many early warning biomonitoring systems. On the other hand, if the purpose of the early warning system is to detect rapidly acutely toxic conditions (such as from industrial spills), then more easily observable response variables such as locomotory behavior may be appropriate.

A trade-off usually exists between the technical ease and cost of measuring a particular organism's response variable and the toxicological sensitivity of that reponse variable. In general, those response variables which are more easily observed usually require less sophisticated equipment to monitor and quantify then those responses which are more subtle. However, the more easily observed responses, as discussed above, are generally less sensitive indicators of toxic conditions than more subtle and less easily observable responses. Therefore, an important conceptual issue in the development of any early warning system is the exact purpose for which the system is designed along with an appropriate compromise of the tradeoffs in cost versus system sensitivity.

The effect of various sensor response variables on overall system sensitivity is well illustrated in an example from our own system development. Earlier versions of our fish ventilatory system (Gruber and Cairns, 1981; Gruber *et al.* 1981) utilized only ventilatory frequency as the sensor response variable. These systems characteristically were able to identify toxic conditions approaching the LC_{50} concentration range (acutely toxic) within 30 minutes. Sub-acutely toxic conditions (20–50% of the

LC_{50} concentration) were detected usually within 24 hours of exposure, and concentrations less than 10% of the LC_{50} concentration were undetectable (Fig. 2.4). Similar results have been obtained by other workers (including several authors

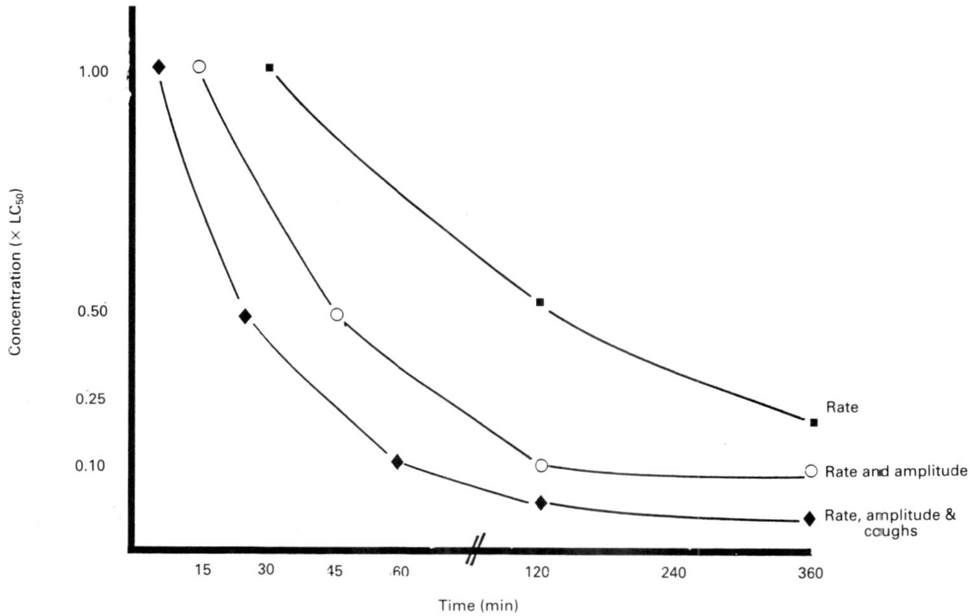

Fig. 2.4 — Sensitivity of a ventilatory early warning system using different complements of fish responses in the analysis.

whose research is presented in this book) using fish ventilatory frequency monitors. Later versions of our system, utilizing frequency and amplitude (depth of breathing) as the sensor response variables (Gruber *et al.*, in press), increased overall system sensitivity markedly, reducing the time to elicit alarms under acute and sub-acutely toxic conditions (Fig. 2.4). Recent developmental systems (Diamond and Gruber, in press) utilizing frequency, amplitude, cough patterns (gill purges), and general activity as the sensor response variables appear to be even more sensitive to sub-acutely toxic conditions (Fig. 2.4). Thus, in our particular system, the more behavioral and biological information we utilized from the sensors, the more sensitive and rapid the response of the system as a whole.

2.6 THE ALGORITHM

If a rapid response (minutes–hours) to acute and sub-acutely toxic conditions is desirable in a given early warning system, then it is usually necessary to incorporate statistical evaluations of the response data into the algorithm or model in the system.

This is because the system by necessity requires the use of comparisons between exposed and unexposed sensors and often small changes in the response variable being measured.

Several factors need to be addressed in order to build a reliable, working algorithm for a given early warning system. Fig. 2.5 illustrates some of these factors.

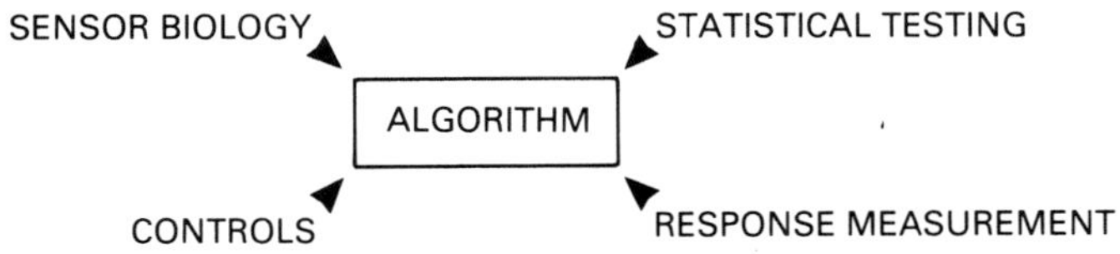

Fig. 2.5 — Components of the algorithm used in BEWS.

By way of explanation, we will use some examples from our own fish ventilatory system in order to illustrate the relationship between these factors and the algorithm.

2.6.1 Sensor biology and algorithm development

Perhaps the most critical aspect of any BEWS is a basic understanding of the sensor behavior and physiology. Insufficient knowledge about sensor behavior and physiology can result in inappropriate response measurements, improper control data, and poor overall system reliability. For example, in our own systems, it was determined that a 48-hour acclimation period (under control conditions) was required prior to any data collection in order to obtain reasonable control data. Without reasonable control data, sensitivity and reliability of the system was compromised since overall variability in control sensor response masked any real changes in response over time due to a pollutant.

Another example showing the importance of sensor biology in designing an algorithm is the occurrence of diurnal rhythms in vertebrate and invertebrate sensor behavior. It has long been known, for example, that fish activity (including ventilatory behavior) follows a predictable diurnal rhythm (Fig. 2.6). This type of information must be incorporated into the algorithm in order to ensure that 'exposed' sensors are evaluated on the basis of a proper control. For this reason, our systems have had to incorporate reference or control values taken continually throughout the day in order to be able to make proper comparisons between exposed and control groups of sensors. A single reference or control value for all time periods would be inappropriate in our system and could result in either excessive false alarms or lack of sensitivity as noted by other workers (Maki, 1979).

2.6.2 Response measurement and evaluation in the algorithm

The way in which sensor responses are compared and evaluated over time in the algorithm can have an enormous effect on system sensitivity and reliability. For example, in earlier versions of our fish ventilatory system, ventilatory frequency in exposed fish at time x was compared with the frequency of those same sensors at time $x-10$ minutes. In theory, changes over time should be detected in this manner and

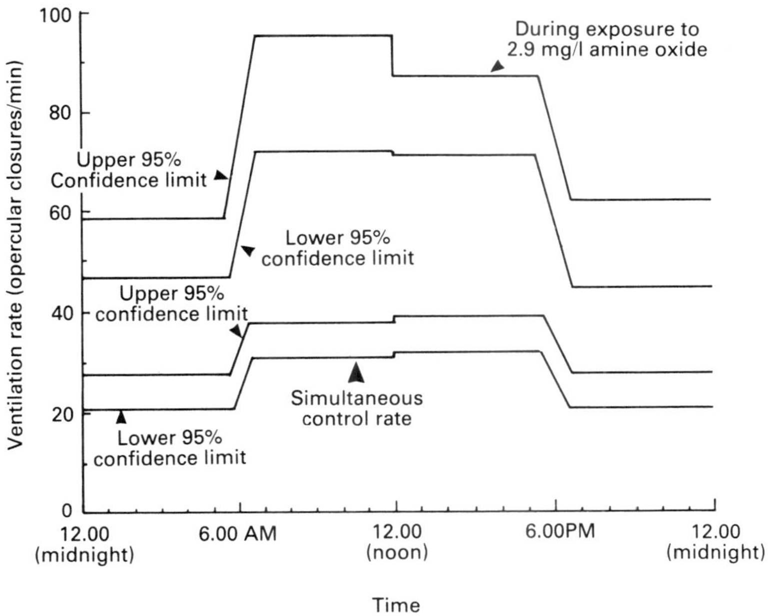

Fig. 2.6 — Fish ventilatory rate upon exposure to amine oxide as a function of time showing the diurnal rhythm of ventilatory rate.

alarms quickly generated if enough sensors responded in a similar manner. In practice, however, the ventilatory response of sensors often changed gradually over time and with considerable variability among sensors resulting in non-significant changes in response over time despite the existence of toxic conditions (Fig. 2.7). The result in these cases was that no alarm would be generated even though over prolonged time periods (hours), a significant change in response occurred. Clearly then, how the algorithm logs and evaluates the response data over time is critical to the overall reliability and sensitivity of the early warning system.

2.6.3 Statistical evaluation techniques utilized in the algorithm

Many early warning systems which have been tested to date have incorporated relatively simple techniques or no statistical testing in the data evaluation process. Instead these systems generally relied on a 'voting' or 'majority rule' concept. That is, an alarm was generated if a predetermined number of sensors responded differently from control sensors in the same time interval. So long as the response measured was fairly obvious (i.e. either present or absent) and high sensitivity to toxicants was not required (i.e. rapid response to acutely toxic conditions was all that was required), the 'majority rule' approach in handling data evaluation is probably appropriate. This technique has the advantages of being relatively simple to incorporate and automate in a system and also tends to reduce the rate at which 'false alarms' occur. The latter is generally true since the 'majority rule' approach tends to

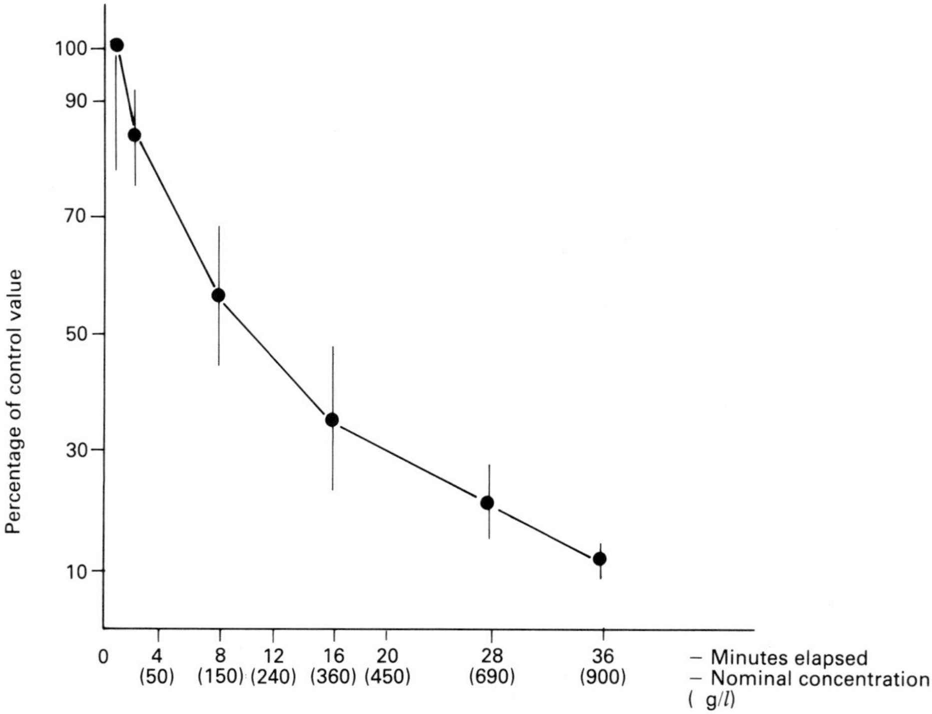

Fig. 2.7 — Ventilatory amplitude of bluegill sunfish plotted as a function of time and increasing zinc sulfate concentration showing gradual decrease in amplitude over time.

provide a conservative estimate of hazard due to the fairly strong response necessary to elicit alarm conditions.

Sophisticated statistical techniques may be necessary in order to evaluate the response data if: (a) the sensor response variable being measured is not an 'all-or-none' response (such as ventilatory frequency, activity, locomotory behavior) and/or (b) a relatively high degree of system sensitivity to pollutants is necessary. In these cases, appropriate statistical procedures can increase the sensitivity of the overall system to lower toxicant levels than might otherwise be obtained. An example of this point can be seen in the historical development of fish ventilatory systems. Response data in earlier prototypes was often analysed using relatively simple comparative methods between exposed and control sensors (Cairns *et al.*, 1970; Slooff, 1979; Sparks *et al.*, 1972; Waller and Cairns, 1972). In many of these studies, data was analysed at preset time intervals with intervening data being discarded. Table 2.1 indicates the exposure time necessary to obtain an alarm state (significant response) to various pollutants in those studies. More recent ventilatory systems utilizing continuous collection of data and a time-series statistical evaluation process decreased the response time for most pollutants (Table 2.1). Thus, system sensitivity was increased by using more of the data generated and by applying more sophisti-

Table 2.1 — Comparison of times to alarm (detection time) in fish ventilatory systems using manual analysis and computer automated analysis

Pollutant	Detection limit (mg/L)	Detection time (min)	
		Manual	Automated
Ammonia (free)	0.11	120	17
Paraquat	1.0	1440	26
Phenol	2.0	660	28
Pentachlorophenol	0.14	720	33
Naphthalene	0.01	950	35
Cadmium	0.025	700	20
Trichlorethylene	0.01	N.D.	40

ND, not detected within a 24-hour monitoring period.

cated statistical procedures to the data obtained. Chapter 4 presents a more thorough discussion of this point and is applicable to any BEWS currently developed.

2.7 FUTURE RESEARCH NEEDS IN THE DEVELOPMENT OF BEWS

2.7.1 No on-line systems to date are foolproof for detecting chronically toxic (sublethal) levels of substances.

Many biomonitoring systems tested (including those discussed in subsequent chapters) have been shown to be fairly reliable and sensitive for detecting acutely toxic (lethal) levels of a wide variety of substances in surface water. Several instances were reported in which some of these systems responded prior to (in some cases several days before) the occurrence of obvious signs of water quality degradation in the environment. Indeed, all of these systems were originally designed with the intention of protecting water quality in the event of 'slugs' or temporary doses of acutely toxic substances in the water column. It should be recognized that this attribute of early warning biomonitors is extremely useful even for drinking water supplies and this should be considered one of BEWS primary contributions.

Since early warning systems have thus far been designed to protect wastewater discharges and other situations where acute toxicity was a potential problem, most of these systems cannot be considered reliable for rapidly detecting chronic levels of toxicants. Given the gradually diminishing supply of good drinking water available in the US and abroad and the increasing number of substances produced which may be toxic or carcinogenic at low concentrations, we feel that early warning systems may be able to play a major role in detecting these problems and guarding against human health hazards.

In our own laboratory, we have recently been experimenting with fish ventilatory-activity systems in order to increase system sensitivity to chronically toxic water conditions. Thus far, our results indicate that chronically toxic concentrations of certain heavy metals and chlorinated hydrocarbons can yield significant responses in under 30 minutes of exposure. Furthermore, there appears to be a consistent type of response in our system depending on the chemical nature of the toxicant. The response to the heavy metals was considerably different from the response to the chlorinated hydrocarbons (Fig. 2.8) suggesting that this type of system may be useful

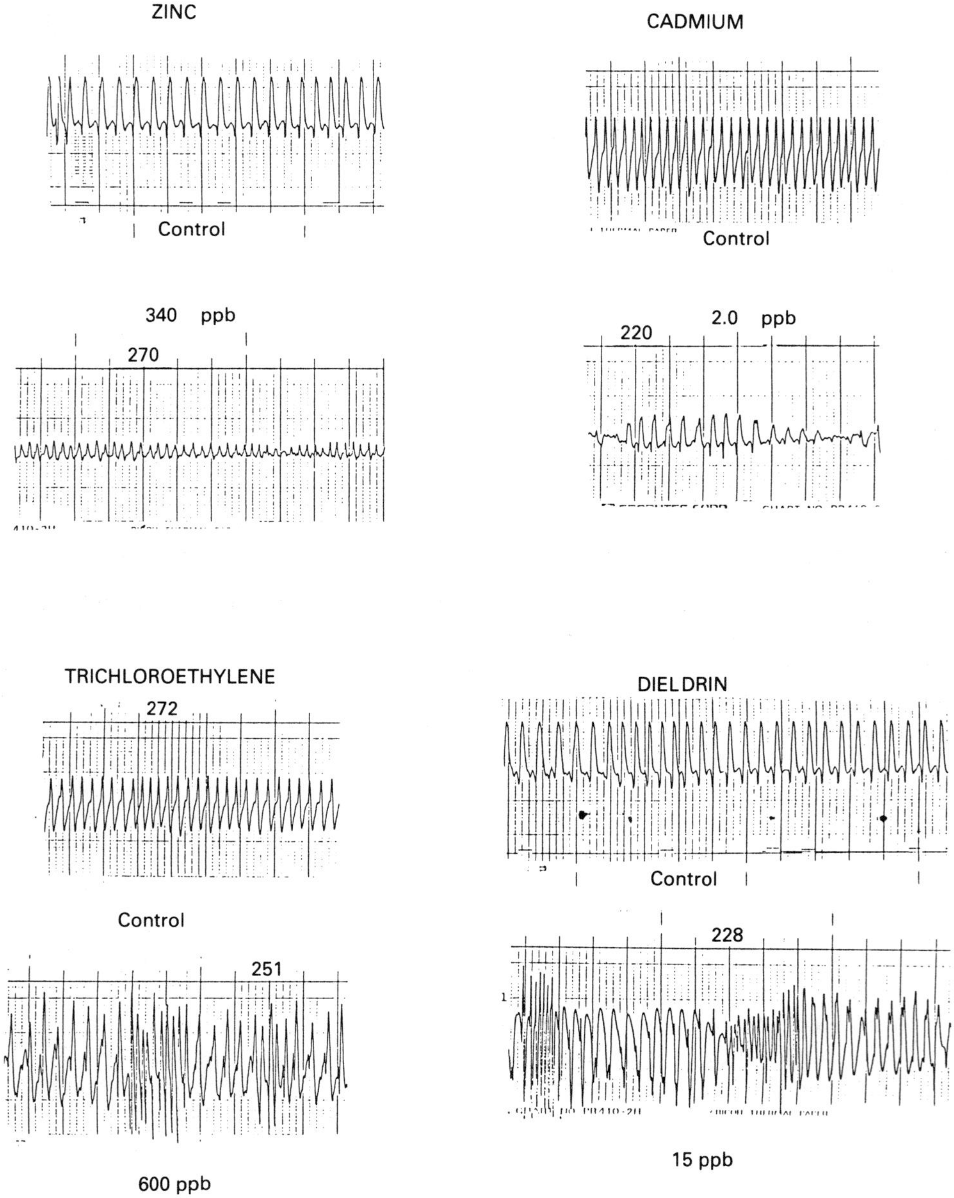

Fig. 2.8 — Ventilatory profiles for bluegill sunfish exposed to control conditions and sub-acute concentrations of various pollutants.

for identifying the nature of the pollutant as well as the existence of chronically toxic conditions. Our preliminary results suggest that with refined data capture and analysis techniques, fish ventilatory systems may be able to detect certain pollutants at levels applicable to present drinking water standards. We feel that this is a productive avenue of research for BEWS development in general, and one which should receive more attention.

2.7.2 No one biological sensor is likely to be sensitive to all pollutants of concern

As shown earlier, it is clear that most biological sensors are very sensitive to the presence of some substances and not others. When dealing with the often wide number of toxicants found in surface waters exposed to industrial effluents and surface runoff, this problem of differential sensitivity can become magnified such that the overall effectiveness of the early warning system is reduced.

If only certain types of toxicants are of special concern, many systems can be tailored to utilize sensors which are especially sensitive to the pollutants of concern. However, as pointed out earlier, sensor selection will necessarily depend on a host of factors which are not directly associated with optimal sensitivity. Such factors include availability, ease of sensor maintenance, and adaptability of the sensor to the overall system environment. Consideration of these factors generally leads to a compromise between sensor sensitivity and sensor reliability.

For this reason we feel that a multisensor system may be a useful approach for early warning monitoring in that such systems should theoretically have increased sensitivity to a wider variety of pollutants than would be possible with one sensor only. Field tests using a multisensor system have been limited thus far so that the reliability and actual sensitivity of such a system cannot be judged at this time. However, such an approach deserves more attention from researchers.

2.7.3 There exists a trade-off between overall system sensitivity to toxicants and system reliability with respect to non-toxic changes in water quality

Most of the response variables and field-tested systems to date have been conservative with respect to overall sensitivity to toxicants. This feature causes these systems to be relatively insensitive to subtle changes in toxicity but enables then to have a relatively low false-alarm rate.

Sensitivity and reliability (under non-harmful conditions) are two parameters which are usually diametrically opposed in an early warning biomonitoring system. If a system is designed to be ultra-sensitive to toxicants, then non-toxic changes in the environment (changes in temperature, turbidity, or conductivity, for example) are likely to be registered by the sensors as well, eliciting a system response. This trade-off becomes particularly crucial when considering early warning systems designed specifically for drinking water supplies. Such a system must, of necessity, be very sensitive to small changes in the concentrations of certain substances, especially those which are potential carcinogens. To date, no system has been field-tested which has shown this type of sensitivity and which also has a low false-alarm rate. In order to guard against a high false-alarm rate in an early warning system it is advisable and probably necessary to have full continuous physicochemical monitoring capabilities simultaneously with the biomonitoring. Many of the biomonitoring systems discussed in this book incorporated such physicochemical (temperature, dissolved oxygen, pH, conductivity, turbidity, etc.) monitoring as well. We see this as essential for two reasons. First, a biomonitoring system may indicate that there is a water quality hazard but it is unlikely to indicate the cause of the hazard. Chemical monitoring will help elucidate the precise nature of the problem. Second, should an alarm be generated in a biomonitoring system, simultaneous chemical monitoring for basic water quality parameters can help pinpoint whether the alarm was caused by a sudden non-toxic change in water quality or whether a potential environmental and health hazard is in the making. As stated earlier, biomonitoring does not replace chemical monitoring in importance. The two components are most effective when used together.

REFERENCES

Cairns Jr, J. (1983) Are single species toxicity tests alone adequate for estimating environmental hazards? *Hydrobiologia* **100** 47–57.

Cairns Jr, J. (1984) Keynote Address. In D. Pascoe and R. Edwards (eds) *Freshwater Biological Monitoring*. Pergamon Press, Oxford, England. pp. 1–14.

Cairns Jr, J. and Dickson, K. (1980) The ABC's of Biological Monitoring. In C. Mocutt and J. Stauffer Jr. (eds) *Biological Monitoring of Fish*. Lexington Books, Lexington, Massachusetts. pp. 1–32.

Cairns Jr, J. and van der Schalie, W. H. (1980) Biological monitoring Part I: Early warning systems. *Water Res.* **14** 1179–1196.

Cairns, Jr, J., Dickson, K. L., Sparks, R. E. and Waller, W. T. (1970) A preliminary report on rapid biological information systems for water pollution control. *J. Wat. Pollut. Control. Fed.* **42** 685–703.

Diamond, J. and Gruber, D. (in press) Detection of chronic toxicity using fish ventilatory behavior. In: *Behavioral Toxicology Symposium*, Society of Environmental Toxicology and Chemistry Special Publication.

Dickson, K. L., Maki, A. W. and Cairns Jr, J. (1979) *Analyzing the hazards evaluation process*. American Fisheries Society, Washington DC.

Ewell, W. S., Gorusch, J. W., Kringle, R. O., Robillard, K. A. and Spiegel, R. C. (1986) Simultaneous evaluation of the acute effects of chemicals on seven aquatic species. *Envir. Tox. Chem.* **5** 831–840.

Gruber, D. and Cairns, Jr, J. (1981) Industrial effluent monitoring incorporating a recent automated fish biomonitoring system. *Water, Air and Soil Pollution* **15** 471.

Gruber, D., Cairns, Jr, J. and Hendricks, A. C. (1981) Computerized biological monitoring demonstrating wastewater discharge compliance. *J. Water Pollution Control Federation* **53** 505.

Gruber, D., Diamond, J. and Johnson, D. (in press). Performance and validation of an on-line fish ventilatory early warning system. In: *Aquatic toxicology and hazard assessment, symposium*. American Society of Testing and Materials, Philadelphia.

Kenaga, E. E. (1978) Test organisms and methods useful for the early assessment of acute toxicity of chemicals. *Environ. Sci. Technol.* **12** 1322.

Kimerle, R. A., Werner, A. F. and Adams, W. J. (1983) Aquatic hazard evaluation principles applied to the development of water quality criteria. In: *Aquatic*

Texocology and Hazard Assessment. STP 854, pp. 538–547. American Society for Testing and Materials, Philadelphia.

Maki, A. W. (1979) Respiratory activity of fish as a predictor of chronic fish toxicity values for surfactants. *Second aquatic toxicology symposium,* STP 607, pp. 77. American Society for Testing and Materials, Philadelphia.

Mehrle, P. and Mayer, F (1985) Biochemistry/Physiology. In: G. Rand and S. Petrocelli (eds) *Fundamentals of Aquatic Toxicology*, pp. 264–282. Memisphere Publishing, New York.

Miller, T., Melancon, S. and Lapoint, T. (1986) Use of effluent toxicity tests in predicting the effect of metals on receiving stream invertebrate communities. In: H. Berman, *et al.* (eds) *Environmental Hazard Assessment of Effluents*, pp. 265–281. Pergamon Press, New York.

Morgan, W. S. and Kühn, P. C. (1984) Aspects of utilizing continuous automatic fish biomonitoring systems for industrial effluent control. In: D. Pascoe and R. Edwards (eds) *Freshwater Biological Monitoring*, pp.65–74. Pergamon Press, Oxford.

Morgan, E. L., Young, R. C. and Crane, C. M. (1984) Automated multispecies biomonitoring employing fish and aquatic invertebrates of various trophic levels. In: D. Pascoe and R. Edwards (eds) *Freshwater Biological Monitoring*, pp. 75–80. Pergamon Press, Oxford.

Slooff, W. (1979) Detection limits of a biological monitoring system based on fish respiration. *Bull. Environ. Contam. Toxicol.* **23** 517.

Slooff, W., van Oers, J. A. and de Zwart, D. (1986) Margins of uncertainty in ecotoxicological hazard assessment. *Envir. Tox. Chem.* **5** 841–852.

Sparks, R. E., Cairns Jr, J. and Heath, A. G. (1972) The use of bluegill breathing rates to detect zinc. *Water Res.* **6** 895.

US EPA (1985a) *Methods for estimating the acute toxicity of effluents to freshwater and marine organisms.* W. Peltier and C. I. Weber. EPA 600/4–85/013, Washington, DC.

US EPA (1985b) *Short-term methods for estimating chronic toxicity of effluent and receiving waters to freshwater organisms.* W. B. Horning II and C. I. Weber. EPA 600/4–85/014, Washington, DC.

US EPA (1985c) *Technical Support Document for Water Quality-based Toxics.* Control Office of Water Enforcement, Washington, DC.

Waller, W. T. and Cairns Jr, J. (1972) The use of fish movement pattern to monitor zinc in water. *Water Res.* **6** 257.

3

Validating biological monitoring

John Cairns Jr
University Center for Environmental and Hazardous Materials Studies, Virginia Polytechnic Institute and State University, Blacksburg, Virginia 24061, USA

3.1 INTRODUCTION

The widely endorsed sequence for hazard assessment of chemical substances in aquatic ecosystems is Tier 1 (screening or rangefinding tests), Tier 2 (predictive toxicity tests), Tier 3 (confirming or validating tests), and Tier 4 (biological monitoring) (Kimerle *et al.*, 1978). In this context, the biological monitoring tier would be on a continuous or systematic basis on natural ecosystems. In a sense, the validating tier would be used to confirm the predictions of the probability of harm at different concentrations of a chemical substance in natural ecosystems based on laboratory tests. Some of the errors likely to occur when using laboratory tests alone have been described by Kimerle (1979). However, the automated biological monitoring systems (ABMSs) developed thus far are primarily early warning systems that have been reviewed by Cairns and van der Schalie (1980). These early warning systems still fit the precise definition of biological monitoring by Hellawell (1978), namely surveillance undertaken to ensure that previously established water quality conditions are being met. Most, but not all (e.g. Morgan *et al.*, 1981) early warning systems, are designed to be put into industrial waste lines to provide a signal of deleterious effects before the waste reaches the receiving system. Cairns *et al.* (1970a,b) describes some of the management options that automated early warning systems provide, based on research begun in 1966. The important difference between automated biological monitoring early warning systems and conventional laboratory toxicity tests (CLTTs) is that the time of information generation in order to take remedial corrective action is being optimized in the early warning systems, whereas precision in establishing a dose/response curve (and thereby being able to estimate the threshold for adverse biological response) is the factor being optimized in the laboratory toxicity tests. Since both are designed to provide information to protect ecosystems, the validation process should be similar in many important respects, but, because of the difference in emphasis, there should also be distinctions in the validation process as well. It is also important to distinguish between the purpose of ABMSs designed for early warning before material reaches the receiving system and that of automated or non-automated biological monitoring systems located in the receiving system itself and designed to provide insurance that the estimates made on the basis of the first three tiers of a hazard assessment process are correct. In this regard, the automated system of Morgan and Eagleson (1983), which is based on the same apparatus used for many early warning systems but located in the receiving system itself and transmitting the results via satellite, should be validated in a fundamentally different way from the automated biological monitoring early warning systems located in industrial manufacturing plants. The purpose of this paper is to examine the correspondence between (a) the response thresholds identified by CLTTs and those identified by automated biological monitoring early warning systems, and also (b) the response thresholds of the indigenous species in the receiving system at different levels of biological organization and the response thresholds of the automated biological monitoring early warning system placed within an industry or other facility discharging waste into a natural system.

3.2 AUTOMATED BIOLOGICAL MONITORING: PROBLEMS OF VALIDATION

The information received from a biological monitoring system as described still needs to be validated in a variety of ways. For some of these, information is already available, and, for others, much additional information needs to be gathered. There is no question but that biological monitoring provides a signal that deleterious biological effects are occurring. Cairns *et al.*, 1977 contains a substantial review of the literature available at that time, particularly on early warning systems together with other forms of biological monitoring. The fact that deleterious biological effects are occurring can be established reliably and statistically, and be suitably replicated. Early warning systems are generally in industrial waste lines or sometimes in water supply systems, and the lag time between the onset of toxic stress and the signal is generally quite small. However, for some substances the lag time could be appreciable. For biological monitoring systems at higher levels of organization than single species, the lag time is substantially greater than the single-species systems that are computer-interfaced, but generally quite rapid. The big question for the 'in-plant' (i.e. industrial manufacturing plant) system is how this signal of deleterious biological effects relates to the response of the natural system. Doane *et al.* (1984) have shown that respiratory rhythm for fish is considerably more sensitive than a variety of commonly used end indicators such as blood chemistry and tissue damage. However, this response has not been adequately related to responses of complex communities naturally occurring in streams, rivers, and lakes. This type of relationship needs to be established unless one assumes that all deleterious biological effects require corrective action even though the sensors may be somewhat more sensitive than the biota in the natural receiving system.

3.3 THE CORRESPONDENCE OF AUTOMATED BIOLOGICAL MONITORING RESULTS TO CONVENTIONAL LABORATORY TOXICITY TESTS

The most straightforward, simplest, and least expensive means of validating the predictions based on automated biological monitoring will be possible if conventional single-species toxicity tests have already been carried out on the same waste discharge or chemical substance and the predictions based on these tests have been validated in natural systems or surrogates thereof (Cairns, 1986, in press). Then, one need merely show that the responses of the automated system can be used to predict accurately the types of responses measured in CLTTs. For example, if one could show in a scientifically justifiable way that alterations of the respiratory rhythm of a large number of individuals of a fish species used as a sensor in an ABMS would, if permitted to continue, result in death, irreversible tissue damage, or significant interference with physiological function, one would then have acceptable validation. This is based on the assumption that (a) these were the end points used in the CLTTs, and (b) these had already been established to have a high correspondence to significant responses in the receiving system such as changes in diversity, nutrient spiralling, or energy flow. One would then have an information system permitting immediate action with confidence when a signal was produced by the ABMS that adverse biological conditions existed.

It is worth re-emphasizing that it is highly improbable that similar end points will be used in the ABMS and the CLTT, the reason being that speed of information generation is being optimized in the former and replicability and reliability as well as accuracy are being emphasized in the latter. Therefore, it should be demonstrated that the same concentration of the chemical that ultimately produces the end points being measured in the CLTT can be reliably predicted from the early warning information generated by the ABMS.

Unfortunately, this ideal situation in which results from the ABMS are validated by comparison with both the results of CLTT and the responses of indigenous biota in the receiving system is not common. However, such extremely useful multiple validations might theoretically be carried out under at least two sets of circumstances: (a) a prudent industry with management that does not like to act on incomplete information and, therefore, has generated a substantial information base in this regard, and (b) a situation where a fishkill has occurred in the vicinity of an industry and that industry is thought to be responsible for it by a regulatory agency, although the industry itself does not feel responsible for the kill. Under these latter circumstances, evidence might be available regarding the concentration of the chemical substance existing in the natural system at the time of the adverse response, and the CLTT would be carried out in the laboratory with the chemical substance to determine if an adverse or deleterious biological response could be obtained at the same concentration or if a much higher concentration were required to do so. Regardless of the outcome of these results, an ABMS might be installed, either to protect the industry from unjust accusations in the future (should the evidence show it was not responsible for the fishkill or other adverse effects in the receiving system) or to alert plant management to the development of toxic conditions in their waste system before the material reached the receiving or natural system.

Regardless of the outcome of this multiple validation process (i.e. either the predictions based on the ABMS were valid or they were not), the information should be published in a professional journal with rigorous peer review. A sufficient number of these multiple validations should establish confidence in the ABMS if the validations are persuasive (i.e. few false negatives or positives) and enable corrections to be made if the correspondence or accuracy of the predictions is not as good as one would hope. Industries would be understandably reluctant to publish this information if it showed that they were unmistakably responsible for the fishkill or other deleterious biological effects in the natural system. However, presumably corrective action would be undertaken in such circumstances, and, if the industry were given appropriate recognition and praise for this corrective action, some of the negative effects of the publication would be mitigated and the positive effects emphasized. In any case, the information is too valuable to the development of ABMS processes to be left unavailable to the larger community of scientists and engineers carrying out research in this field or to be placed in the limited access 'gray literature' lacking the rigorous peer review that establishes credibility of any type of evidence. Therefore, both regulatory agencies and journals should give this problem serious consideration so that this valuable information is made available to the larger community of professionals in the field without further damage to the reputation of the industry that permits this.

3.4 COMPARISON OF ABMS WITH CLTT ONLY

Probably, the most common situation where this will be done will be when CLTTs have been carried out and show that under certain circumstances episodic pulses of biologically deleterious materials appear in the waste discharges of a particular industry. If these circumstances are so complex that they cannot be easily predicted from readily available production/operational evidence, an ABMS might well be installed to alert management to the development of deleterious concentrations of waste materials so that immediate corrective action can be taken before the material reaches the receiving system. Under these circumstances, a management decision might be made to restrict the validation process to confirming that the ABMS can be used to predict accurately the onset of toxic conditions as demonstrated by CLTTs. Since the predictions developed from CLTTs of the responses of complex natural systems have been inadequately validated (Cairns, 1986 and submitted), this represents a faith in the predictive capabilities of CLTTs yet to be thoroughly documented in a rigorously peer-reviewed scientific fashion. Nevertheless, it is the minimal validation that should be carried out in order to make an informed management decision. In short, if the CLTTs are accepted as the sole evidence required by regulatory agencies and the correspondence between the ABMS results and the CLTT results is persuasive, management might make a decision to go only this far, despite the fact that rigorous scientific validation requires that they go further. Although this might seem to be a cost-cutting measure to management (i.e. restricting the validation process), Kimerle *et al.* (1978) have illustrated quite persuasively how conventional clean water laboratory toxicity tests as well as clean water estimates of environmental fate can differ substantially from responses in the more complex 'real world.' In one of the figures in this excellent discussion (Kimerle *et al.,* 1978), the clean water laboratory test showed that the environmental concentration of the chemical and the concentration producing no adverse biological effects were quite close together, although in the real world they were very far apart, the environmental concentration of the chemical being well below the concentration producing adverse biological effects in the real world. Under these circumstances, management might understandably be persuaded to install a more comprehensive waste treatment system, markedly reducing the concentrations of the chemical substances perceived to be deleterious on the basis of the clean water laboratory tests. However, they would be shown not to be deleterious if real world evidence were available. This improved waste treatment system would unquestionably be more expensive both in capital investment and in operating costs than the existing system, but would produce no demonstrable biological benefits. Although no comparison has been made of the cost of generating both clean-water laboratory and real-world evidence to document this, few people would argue that the cost of doing so would be anything more than a fraction of the combined capital investment and operating costs of a sophisticated waste treatment system. In this scenario, the money saved by not carrying out a thorough validation of the CLTTs in 'real-world' systems was spent many times over in installing a waste treatment system that produced no easily demonstrable biological benefits. I have deliberately used the terms *clean-water laboratory tests* and *real world* because they are used in the Kimerle *et al.* (1978) discussion and because they vividly portray the circumstances that cause errors in prediction and, therefore, the results should be suspect until validated.

The alternative scenario depicted in the Kimerle *et al.* (1978) discussion could be equally financially damaging as a consequence of failing to validate thoroughly the predictions based on clean water laboratory tests. In this the scenario, the clean water laboratory tests show that the environmental concentration of the chemical and the concentration of the chemical producing adverse biological effects are very far apart. However, in the 'real-world', the environmental concentration of the chemical and the concentration producing adverse biological response are quite close together. Under these circumstances, a fishkill or other environmental catastrophe might easily occur because changes in environmental conditions could alter both the environmental concentration of the chemical and the threshold producing adverse biological responses in such a way that the environmental concentration of the chemical was well above the concentration producing adverse biological responses. If an attempt had not been made to validate the clean-water laboratory test and the predicitions based on them had not been shown to be inaccurate in the real world, management would have been totally unprepared for this catastrophe. Anyone examining fines levied these days for environmental degradation would almost certainly be persuaded that the cost of validation, which would have lead to detection of this error, would be far less expensive than paying a substantial fine, not to mention the adverse public relations that would be likely to develop from such a scenario. It is especially important to note that in this scenario the results obtained with an ABMS might validate quite well the results in a CLTT and that both might be inaccurate when predicting real-world events. Thus, although it might be tempting to restrict the validation process to obtaining corresponding results with ABMS and

CLTTs, this is not a prudent course of action as long as both might be in error in predicting real-world events. Until a substantial body of information has been generated to indicate the probable accuracy of predictions based on either ABMS or CLTT, it would be prudent to include validation in natural systems, however tempting it might be not to do so. The financial and public relations consequences of making management decisions based on an incomplete validation process are too severe to justify this practice.

3.5 ABMS VERSUS FIELD RESPONSES

It is difficult to imagine a situation in which an ABMS was installed without carrying out at least a limited array of CLTTs. However, it is at least conceivable that some organizations might wish to do so. The most economical way to carry out such evaluations of an ABMS is to determine responses for an array of conditions by means of simulations of a variety of both concentrations and qualitative differences in the industrial waste line, however improbable some of these might be. The ABMS is specifically designed to produce an early warning signal that some unexpected condition has developed, and, therefore, infrequent but still probable circumstances should be simulated. As is customary in rangefinding tests, one need not simulate every condition, but should at least simulate extreme conditions, normal conditions, and several points in-between. In this way, the difference between normal operating conditions and those creating a signal for deleterious response can be determined with reasonable reliability. Since one cannot and should not use natural systems for a rangefinding validation, a suitable surrogate such as a microcosm or mesocosm should be developed for this purpose. The microcosm could be as simple as the one described by Cairns *et al.* (1986) and Cairns and Pratt (1985) or as elaborate as some of those described by Odum (1984). The important determination is whether the ABMS can reliably predict responses in more complex, environmentally realistic systems acting as surrogates of natural systems. This form of validation would enable one to decrease the number of false positives and false negatives generated by the ABMS.

One might further validate this system when waste discharges are going into a receiving system by periodic field surveys in the natural system to confirm that when the ABMS is indicating no adverse effects the condition of the natural system verifies this prediction. One might even use a field enclosure (e.g. Odum, 1984; National Research Council, 1981) to validate further the predictive capabilities of the ABMS by creating situations in the field enclosure comparable in concentration to those of the chemical substance that produce an adverse biological effects signal in the ABMS.

3.6 HOW MUCH VALIDATION IS ENOUGH?

It seems quite clear that there should be at least two phases in developing a scientifically justifiable, cost-effective validation process. Phase 1 would be the research phase in which the overall relationship between the ABMS responses and the CLTT responses are compared with each other and ultimately both would be compared with the responses in either natural systems or surrogates of natural systems such as microcosms and mesocosms. This does not mean endless study but rather that a substantial evidence base of these relationships must be established to reduce the uncertainty in them. Almost certainly this process will be accelerated if industries engaging in generating such data, which ultimately become available for general use, are given some incentives to do so. Since tax writeoffs are, at least temporarily, less popular than they were a few years ago, perhaps some regulatory incentives could be developed, such as a variance or waiver permitting discharges at any rate or amount for which there is no compelling evidence of adverse biological effects in the receiving system. If the correspondence in the thresholds in the three different systems (i.e. ABMS, CLTT, natural systems) is quite good, this period of development for an information base should be relatively brief. If the correspondence is not good, the source of the errors in prediction will need to be identified and corrected. It is exceedingly difficult to predict the length of this developmental phase, the size of the information base required, or the degree of correspondence in response between the various thresholds. However long this process takes, it is quite clear that automated biological monitoring cannot develop much further without it. The major value of automated biological monitoring is in providing a reliable, economically feasible, rapid biological information system to give an early warning of impending harm. Until full confidence is established in this system as a consequence of validation, the information generated in this process should tell us a great deal about the value of information generated in all three systems being compared.

The second phase of development will be relatively straightforward. Once the general relationships are established, the information generated will only need to confirm on a site-specific basis that the generalizations already established are true for this site or, if they are not, to determine the factors that make them different. At this particular point in development, the information will have less value per unit generated than it did earlier during the period of great uncertainty, but it will nevertheless be quite valuable for a number of years. This will be especially true if at certain sites the generally useful models do not work.

After phase 2 is well underway, it would seem reasonable to modify existing legislation to encourage the replacement of arbitrary standards or less sophisticated methods with this particular approach. One of the incentives for installing and maintaining an ABMS should be the ability of organizations possessing such instrumentation to come closer to fully utilizing the assimilative capacity of natural systems by permitting them to come closer to response thresholds than regulations based on worst possible case situations and substantial application factors permit them to do.

ACKNOWLEDGMENTS

This manuscript was prepared entirely at Rocky Mountain Biological Laboratory during summer, 1986. I am indebted to Darla Donald, Editorial Assistant, University Center for Environmental and Hazardous Materials Studies for preparing this manuscript for publication, and to Becky Allen for typing the rough draft and Betty Higginbotham for typing the final draft.

REFERENCES

Cairns Jr, J. (1986) What is meant by validation of predictions based on laboratory toxicity tests? *Hydrobiologia* **137** 271–278.

Cairns Jr, J. (in press) What constitutes field validation of predictions based on laboratory evidence? In R. J. Livingston (ed.) *Biological assessment of hazardous wastes.* American Society for Testing and Materials, Philadelphia.

Cairns Jr, J. (submitted) Keynote address: Politics, economics, science — going beyond disciplinary boundaries to protect aquatic ecosystems. In J. Gannon and M. Evans (eds) *Persistent toxic substances and the health of aquatic communities.* John Wiley, New York.

Cairns Jr, J. and Pratt, J. R. (1985) Multispecies toxicity testing using indigenous organisms — a new cost-effective approach to ecosystem protection. pp 149–159 in 1985 Environmental Conference, ISSN 0272–7269. Tappi Press, Atlanta.

Cairns Jr, J. and van der Schalie, W. H. (1980) Biological monitoring. Part I: Early warning systems. *Water Res.* **14** 1179–1196.

Cairns Jr, J., Dickson, K. L., Sparks, R. E., and Waller, W. T. (1970a) A preliminary report on rapid biological information systems for water pollution control. *J. Water Pollut. Control Fed.* **42**(5) 685–703.

Cairns Jr, J., Dickson, K. L., Sparks, R. E., and Waller, W. T. (1970b) Reducing the time lag in biological information systems. 159th Annual Meeting, American Chemical Society, Division of Water, Air and Waste Chemistry. **10**(1) 84–98.

Cairns Jr, J., Dickson, K. L. and Westlake, G. F. (eds) (1977) *Biological monitoring of water and effluent quality.* STP607. American Society for Testing and Materials, Philadelphia. 246 pp.

Cairns Jr, J., Pratt, J. R., Niederlehner, B. R., and McCormick, P. V. (1986) A simple, cost-effective multispecies toxicity test using organisms with a cosmopolitan distribution. *Environ. Monit. Assess.* **6** 207–220.

Doane, T. R., Cairns Jr, J., and Buikema Jr, A. L. (1984) Comparison of biomonitoring techniques for evaluating effects of jet fuel on bluegill sunfish (*Lepomis macrochirus*). In D. Pascoe and R. W. Edwards (eds) *Freshwater Biological Monitoring,* pp. 103–112 Pergamon Press, Oxford.

Hellawell, J. M. (1978) *Biological Surveillance of Rivers.* Water Research Centre, Stevenage, England. 332 pp.

Kimerle, R. A. (1979) Aquatic hazard assessment: concepts and applications. In M. P. Bratzel (ed.) *Workshop on Hazard Assessment,* pp 221–230. Great Lakes Water Quality Board of the International Joint Commission, Windsor, Ontario.

Kimerle, R. A., Gledhill, W. E., and Levinskas, G. J. (1978) Environmental safety assessment of new materials. In J. Cairns Jr, K. L. Dickson and A. W. Maki (eds) *Estimating the Hazard of Chemical Substances to Aquatic Life,* STP657, pp 132–146. American Society for Testing and Materials, Philadelphia.

Morgan, E. L. and Eagleson, K. W. (1983) Remote biosensing employing fish as real time monitors of water quality events. In *Hydrological Applications of Remote Sensing and Remote Data Transmission* (Proceedings of the Hamburg Symposium, August, 1983). IAHS Publ. No. 145, pp 235–243.

Morgan, E. L., Eagleson, K. W., Hermann, R., and McCollough, N. D. (1981) New developments in automated biosensing from remote water quality stations and

satellite data retrieval for resources management. *J. Hydrol.* **51** 339–345.
National Research Council (1981) *Testing for Effects of Chemicals on Ecosystems.* National Academy Press, Washington DC. 103 pp.
Odum, E. P. (1984) The mesocosm. *BioScience* **34**(9) 558–562.

4

Altered ventilatory frequency from fish in an automated biomonitoring system: how to evaluate if it is significant

Robert S. Hayward, Norman G. Reichenbach and **L. Barry Goss**
Environmental Sciences Department, Battelle Columbus Division Columbus, Ohio, USA.
Lula A. Dickson and **Thomas J. Wildoner Jr**
Pine Bluff Arsenal, Pine Bluff, Arkansas, USA.

4.1 INTRODUCTION

Studies involving various fish species have shown that changes in their ventilatory frequencies (VFs) away from some basal pattern may be expected when these animals are stressed (Cairns *et al.*, 1980; Cairns and Garton 1982). The nature of ventilatory rate responses of various fishes to known toxicants, delivered as single or incremented dosages, has been studied in laboratory settings. This, in combination with the fact that the bluegill, *Lepomis macrochirus,* is one of the more highly studied indicator fish species in these settings, suggests that the development of a continuous biomonitoring system involving this species would be a straightforward transition, once hardware-related problems have been resolved. In fact, a suite of new issues of analytical, physiological, and behavioral natures are peculiar to the continuous biomonitoring setting and deserve, if not require, attention if true alarm situations are going to be properly assessed. Many of these new issues arise from the inherent variabilities associated with time. Supporting laboratory studies are typically conducted over discrete periods, usually no longer than a few weeks. In contrast, automated biomonitoring systems (ABMS) are continuous, inviting stimuli and responses that may not be present in short-term studies. An example of a time-related issue associated with the continuous ABMS setting involves the maintenance of a consistent, high-quality control environment, against which altered ventilatory responses may be measured. Additionally, fish held for long periods must be fed, which in turn, elevates respiratory rates for periods that are dependent on individual size and ambient temperature. An ability to deal with changes in VFs associated with feeding is necessary. Another possible time-related issue arises from unknown effects of long-term confinement of indicator animals on ventilatory patterns. Changes in basal ventilatory patterns may also occur over longer periods. Just how fish size, sex, and genetic composition may bear on the 'shelf-life' of an indicator animal may be important in future refinements of continuous monitoring systems.

Another set of issues arises from the 'online' nature of the ABMS setting in contrast to the artificial (laboratory) setting. In the online setting, a complex set of toxicants (including, for example, synergisms) is expected to be detected by the ABMS, in contrast to the finite set of usually single toxicants that have been applied in known dosages in most laboratory studies to date. Response patterns of the animals may vary substantially and can include threshold, depressed, or subtly changing responses. Hence, a robustness must be inherent in any analytical approach for detecting altered ventilatory patterns.

The motivation for reconsideration of the alarm evaluation system currently used on the ABMS, a system created by Battelle in conjuction with Pine Bluff Arsenal to monitor, using bluegill fish, the final effluent from a military process wastewater treatment facility, was an inordinately high frequency of 'alarm' reponses registered by the ABMS. The original procedure developed standard 95% confidence intervals for VFs of individual bluegills from a forward-moving set of historical VF data. A VF value falling outside of these limits was taken to indicate a significant response to toxicity. However, ventilatory rates of bluegills are known to be serially correlated (Thomspon *et al.*, 1982), meaning that the respiratory rate at a given point in time is dependent on previous value(s). In general, when data are correlated in this way, the confidence limits calculated are narrower than those when the data are completely independent. Hence, the confidence limits were too narrow (because the data which were used to calculate the confidence intervals were correlated) and these confidence intervals resulted in numerous 'false alarms'.

If the nature of the serial correlation between data for each time interval can be defined, however, one can use this advantageously to predict more precisely where the next point (or points) will fall and how a confidence interval could be correctly calculated.

We considered two fundamentally distinct approaches to the analysis of ventilatory rates in the ABMS setting which would handle the serially correlated nature of the data and accommodate different reponse patterns.

4.2 APPROACH

4.2.1 Individual-based approach

One approach that we considered focuses on the response of the individual fish and is referred to as the 'individual-based approach'. Here we rely on a predictive technique known as time series analysis, the time series under analysis in this case being the sequences, through time, of bluegill ventilatory rates. Time series analysis first identifies the pattern of correlation within a series of data. After defining this correlation, values at future points in time are predicted from values that have already occurred. Confidence intervals based on this analysis are developed which are expected to capature the future values at a given level of probability. These projected limits of future values of VFs from individual fish represent the cut-off values for deciding when fish are showing a response to toxicity.

4.2.2 Group-based approach

An alternative approach to analyzing the VF responses, which also handles the serially correlated nature of the data, considers group-based responses, as opposed to individual-based responses. Three groups of four fish are used, one comprises bluegills each of whose chamber receives a flow of dechlorinated tap water (control), another where each receives a mixture of 50% tap and 50% effluent water, and a third group where each receives 100% effluent water. Each group consists of four fish, each in its own exposure tank. The expectations are that any trends in VFs that are not induced by effluent toxicity should also be present in the tap water group's response. For example, if bluegill VFs were increasing from 0700h through 1000h in response to light intensity, this pattern would be expected to occur in the control group as well as in the two treatment groups (those with effluent water). Differences in the average responses of the three groups should not change from this short-term elevation of the VFs because the three groups are expected to respond similarly. However, when the effluent water contains a toxic component, a change in the VFs for the fish in the 50% and 100% groups are expected, whereas changes in the VFs for the fish in the control group would not occur. Thus, this group-based approach considers differences in the average responses of the three groups. The occurrence of differences between the groups, as detected by simple nonparametric statistical procedures, is used to decide when changes in VFs are indicating the presence of toxic components in the effluent.

4.2.3 The data base

Although we possessed certain intuitive insights into how well specific analyses of the two types (individual-based and group-based) should perform, we assessed comparative performance of selected procedures by applying these to data collected from the ABMS in 1985. We selected ABMS data from two periods, one when VF response patterns were somewhat variable but toxicity in the effluent was not indicated (March 9–14, 1985), and a second (April 25–26, 1985) when a clear toxic response was shown by fish in effluent water (Figs 4.1 and 4.2). We compared responses among analytical techniques to both the more variable data series (March 9–14) and the known toxic response data series (April 25–26).

4.3 APPLICATION AND DISCUSSION

4.3.1 Application of time series analysis (individual-based approach)

A preliminary step in applying time series analysis is the identification of a model that adequately describes the nature of correlation among values in the time series under consideration. One type of model, the one to which we ultimately required an indicator fish's VF pattern to conform before it was used in our analysis, is termed an autoregressive model (AR). Simply instead, this model gives point estimates and confidence limits for future values of a time series using the mean of an observed series, plus information from values at recent points in time. Once such a model is

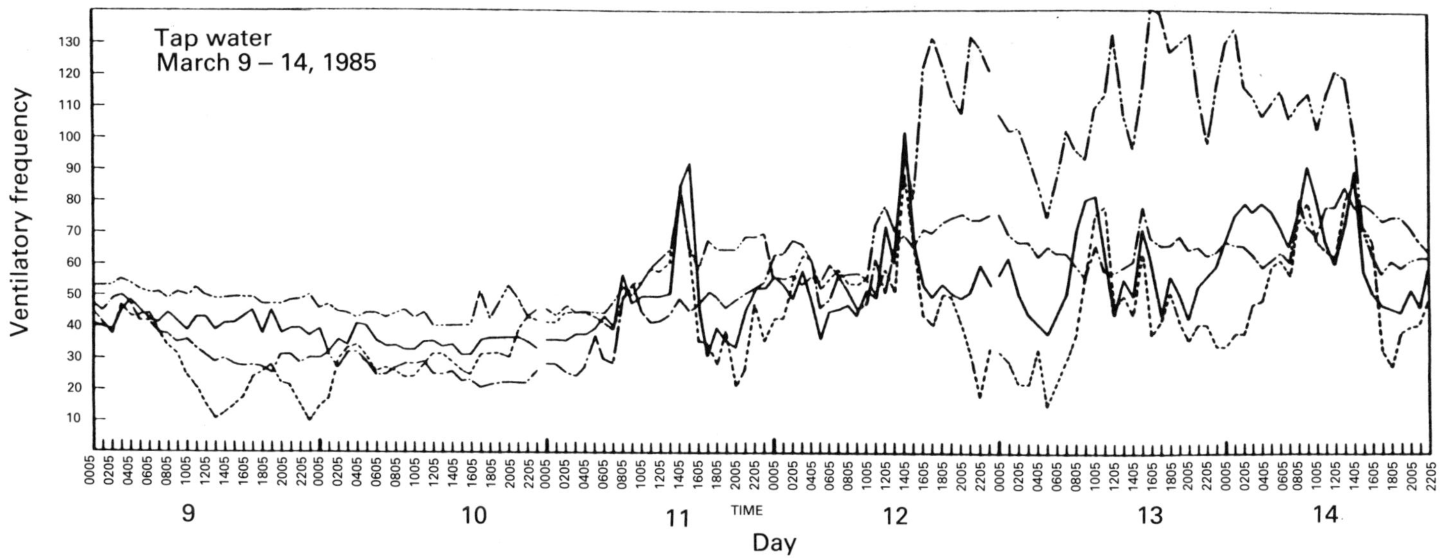

Fig. 4.1 — (a) Ventilatory frequency response patterns of four bluegills in tap water during March 9–14, 1985 when toxicity in the Pine Bluff Arsenal (PBA) effluent (NPDES 002) was considered negligible (Bean, 1985).

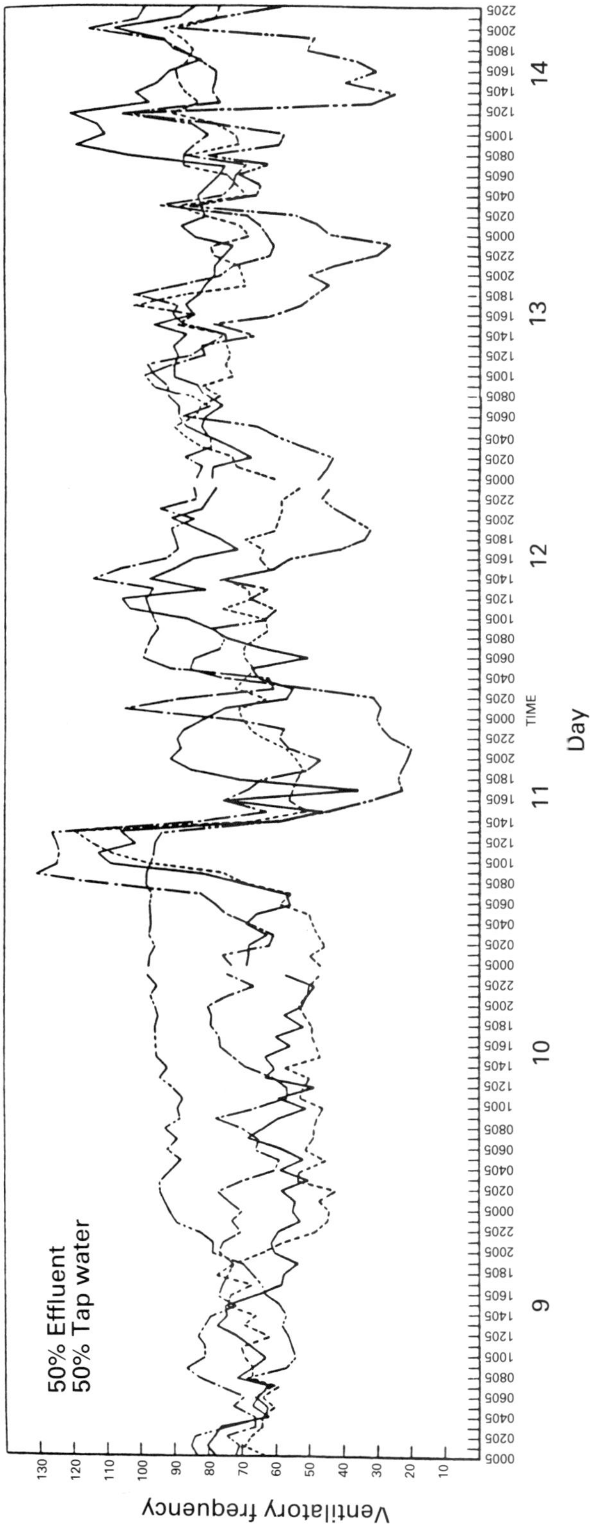

Fig. 4.1 — (b) Ventilatory frequency response patterns of four bluegills in 50% effluent water during March 9–14, 1985 when toxicity in the PBA effluent (NPDES 002) was considered negligible (Bean, 1985).

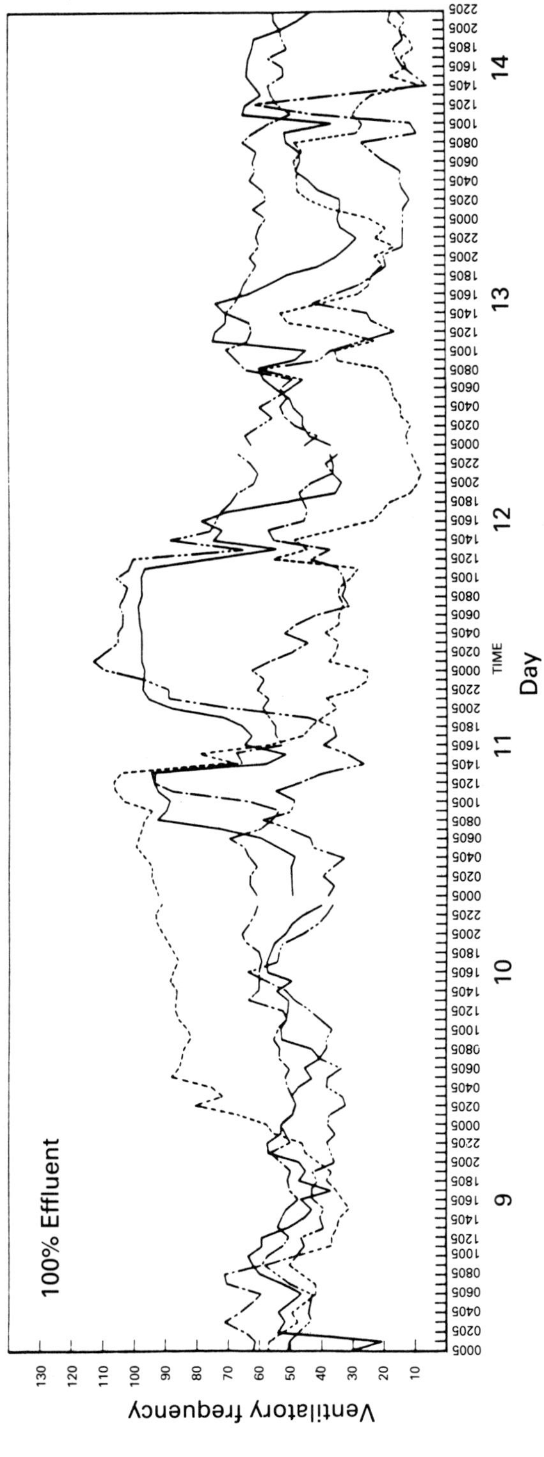

Fig. 4.1 — (c) Ventilatory frequency response patterns of four bluegills in 100% effluent water during March 9–14, 1985 when toxicity in the PBA effluent (NPDES 002) was considered negligible (Bean, 1985).

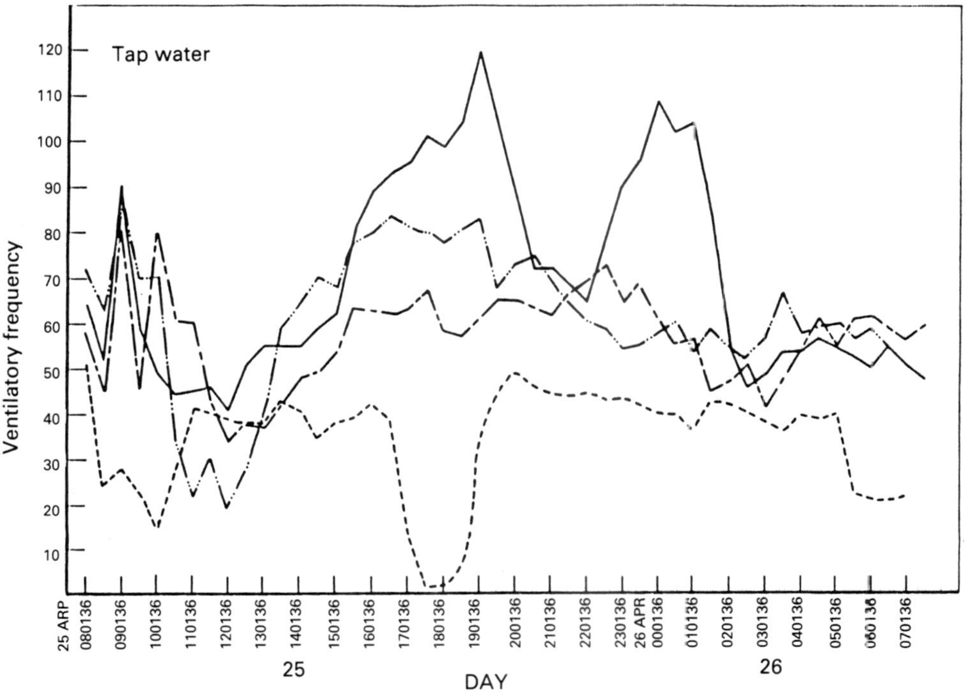

Fig. 2.2 — (a) Ventilatory frequency response pattern of four ABS bluegills in tap water during April 25–26, 1985 when toxic conditions were considered to be present in the PBA effluent (NPDES 002) (Bean, 1985).

selected, its 'order' must be determined. The order of the model simply refers to the number of recent values that will be included in a future projection of a fishes VF. The simplest autoregressive model has an order of one (AR(1)). In this case a future VF value will be estimated from information on the mean of the entire time series data set and the most recent value. An appealing feature of the AR models for our purposes, is that confidence limits for all future values tend to stabilize to a constant range. Hence, no matter how far ahead we wish to project, we expect that future values will fall within the same boundaries. Other models tend to tell us less about future values, since confidence limits usually expand the further ahead in time we project.

To diagnose the appropriate model and order for a given time series, a 'correlogram' is generated using the Statistical Analysis System (Helwig and Council, 1982) (SAS) to show the structure of the correlation between values through time. Thompson *et al.* (1982) evaluated time series of VFs of unstressed bluegills and found that many individuals fit the simple AR(1) model. Other individual bluegill VFs, however, were not adequately described by this simple model. We analyzed the time series of ventilatory rates for bluegills within periods when stress levels were considered to be minimal. We found that a portion of the bluegills did show conformation to the simple AR(1) model. Again, we decided to work further with

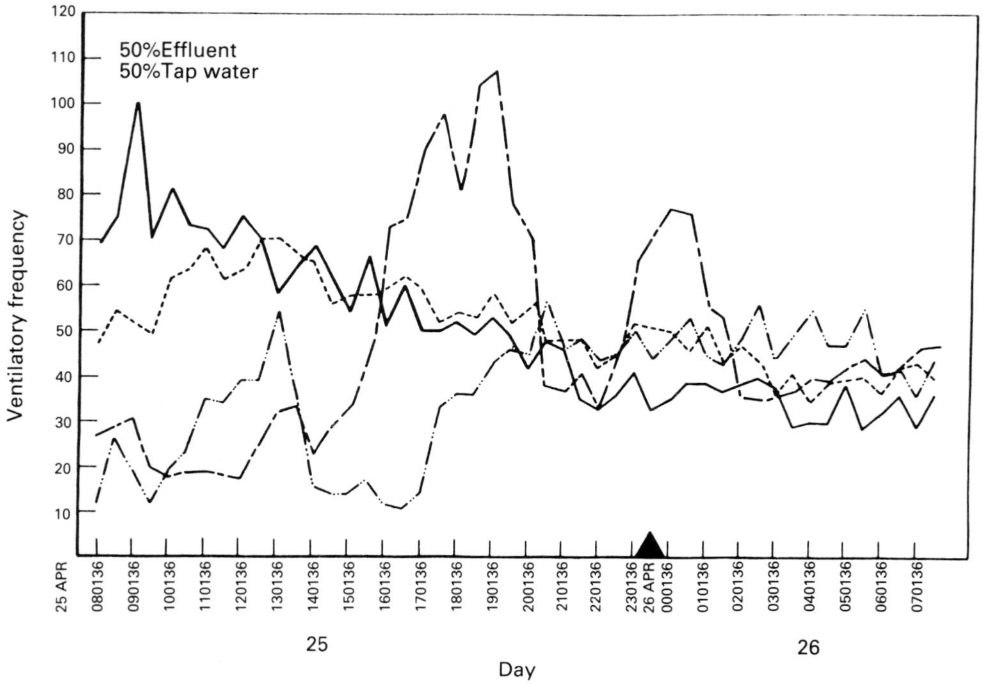

Fig. 2.2 — (b) Ventilatory frequency response pattern of four ABS bluegills in 50% effluent water during April 25–26, 1985 when toxic conditions were considered to be present in the PBA effluent (NPDES 002) (Bean, 1985).

only those fish that conformed to this simple model because long-term forward projections could be more reasonably developed for this set of individuals.

After selecting bluegills whose basal VFs conformed to the simple AR(1) model, time series analysis was applied to the ABMS data. Using a specialized package for time series analysis on SAS (PROC ARIMA), we determined confidence limits (95% probability) for future values of VFs for individual fish that resided in chambers receiving 100% effluent water. We imposed a subjective criterion upon the time series analysis by requiring that three of the four bluegills transcend the confidence limit for a given time interval before significant toxicity was considered as being indicated (Fig. 4.3). Requiring that three of four fish fall beyond these limits in any given hour before a significant response is indicated reduces the likelihood of a false alarm.

4.3.2 Application of group-based approaches

Two types of group-based approach were developed and evaluated using the ABMS data. The first type simply considered differences in median VFs among the tap, 50%, and 100% groups for each hour (absolute method, Fig. 4.4). A second type of group-based approach was based on the differences in a fish's VF in two consecutive hours (delta method, Fig. 4.5). For both types of group-based approach, the median differences for the four fish within a group were compared with the other two groups.

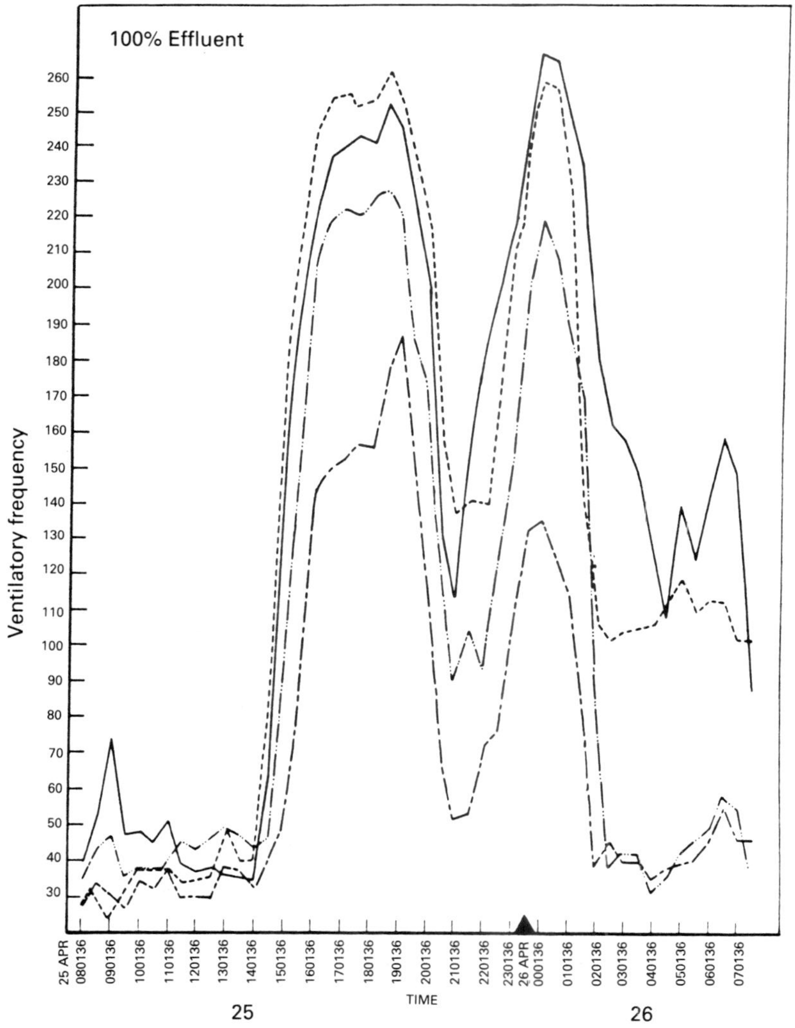

Fig. 2.2 — (c) Ventilatory frequency response pattern of four ABS bluegills in 100% effluent water during April 25–26, 1985 when toxic conditions were considered to be present in the PBA effluent (NPDES 002) (Bean, 1985).

Nine response patterns of median VFs among the three groups could possibly occur for both the absolute and delta group-based methods. Assuming that no toxic effects ever occurred in fish in the tap water group — true control response — six patterns could conceivably arise when significant toxicity is present in the effluent water. Two other response patterns, the 'V', and the inverted 'V', should not occur unless control assumptions are being violated. The last possible pattern, the 'flat' response, would indicate no between-group differences. Explanation for each of the nine response patterns is given in Fig. 4.6. A suite of simple nonparametric statistical

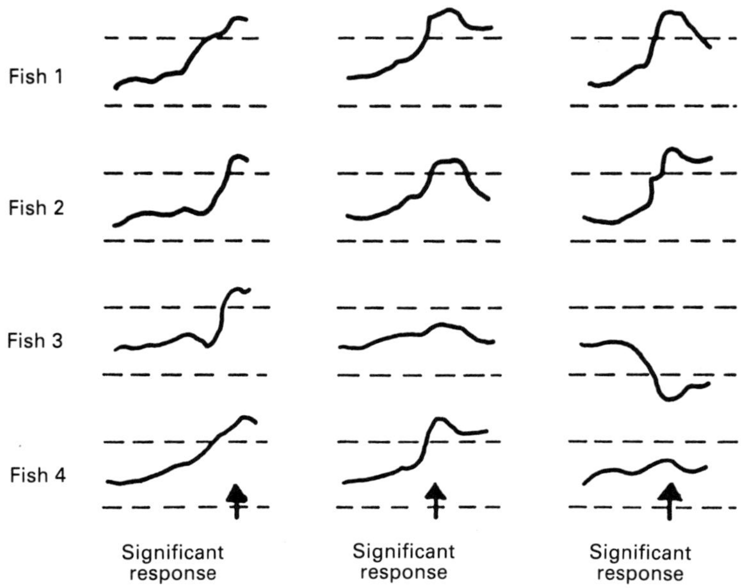

Fig. 4.3 — Demonstration of three possible responses by four bluegills in 100% effluent water, each of which has had 95% confidence limits developed for their individual VF patterns from time series analysis. Note that values of the bounds would differ for each fish. Using the subjective requirement that three of four bluegills show VF values outside of their 95% confidence limits to indicate a significant departure from the basal pattern, all three situations constitute a significant response.

procedures based on ranks were considered for the group-based evaluations. Two-group tests that were considered (where control-50%, 50%–100%, and control–100% were evaluated for differences in their median VFs) included the Mann–Whitney U test and the Fligner–Policello test. Three-group tests (where differences between the three groups would be evaluated simultaneously) included the Kruskal–Wallis test, Jonckheere's test and various 'umbrella' tests (Hollander and Wolfe, 1973). Although the three-group tests are appealing because all three groups can be handled together in a single analysis, we decided to use the Fligner–Policello two-group test in both the absolute and delta procedures since this was the only test, of those considered, which did not assume that the variances for all three groups were homogenous. Our data indicate that the variation of VFs of bluegills within the treatment (effluent) groups may change as toxicity increases and consequently would not be similar to the variation in the VFs for the control group. Hence, in our evaluation of performance of the absolute group-based method, three Fligner–Policello tests were conducted for each hour to determine the response pattern. The objective was to determine whether one of the six toxicity reponse patterns was occurring or whether a flat pattern (indicating no toxicity effect) was present. Detection of either of the two 'V' patterns would indicate problems with the control assumptions (Fig. 4.6).

We developed the delta method as an alternative to the absolute procedure

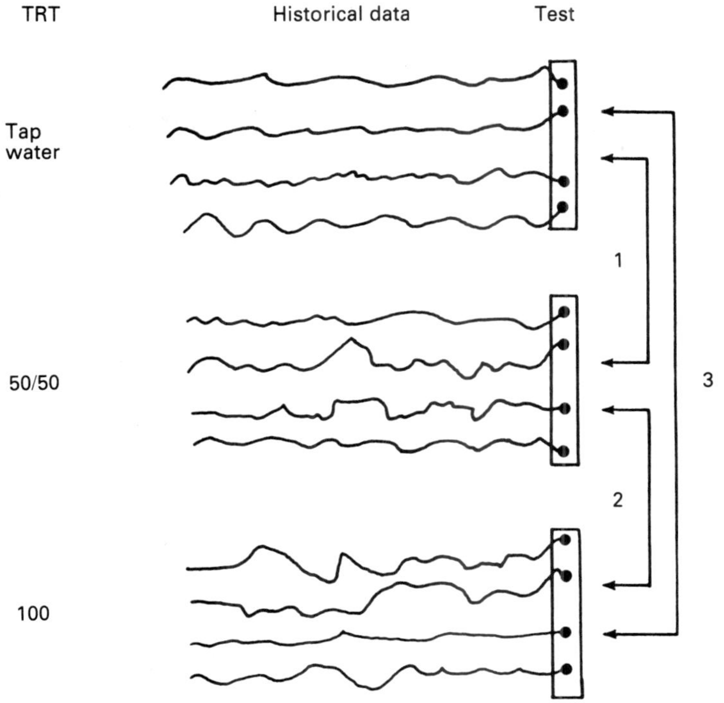

Fig. 4.4 — Diagrammatic representation of the 'absolute' group-based method for evaluating when significant VF responses in the 50% and 100% effluent groups are occurring. The three pairwise comparisons are noted with numerical values.

because it tends to factor out differences between individuals, while picking up only major shifts in the relationships between groups. This procedure was expected to perform well whenever toxicity in the effluent caused pronounced responses (shifts) in the bluegills and it was also expected to have a lower frequency of 'false alarms'. Another positive feature of this method should be its ability to accommodate 'drifting' of the VF series as might occur if, for example, a gradual acclimation response was in progress. Unfortunately, the same kind of gradual response could occur because of toxicity in the effluent and these types of responses, as noted above, would not be readily detected, a potentially undesirable trait.

A comparison of the Fligner–Policello results for the absolute and delta methods when each was applied to the same series of ABMS data is given in Fig. 4.7. Significant intergroup differences are indicated by the absolute method throughout a 12-hour period, followed by a shift to no intergroup differences around 1300h. As expected, the delta method registers a significant response only at the hour of the 'shift' in group difference. Note that this shift by the absolute method at 1300h is towards greater similarity between the tap and 50% group, representing a backshift to nonalarm conditions. The delta method recognizes the shift only, and not its direction. Hence, when using the delta method, all significant responses must be followed by a check to determine whether an intergroup shift(s) has been away from,

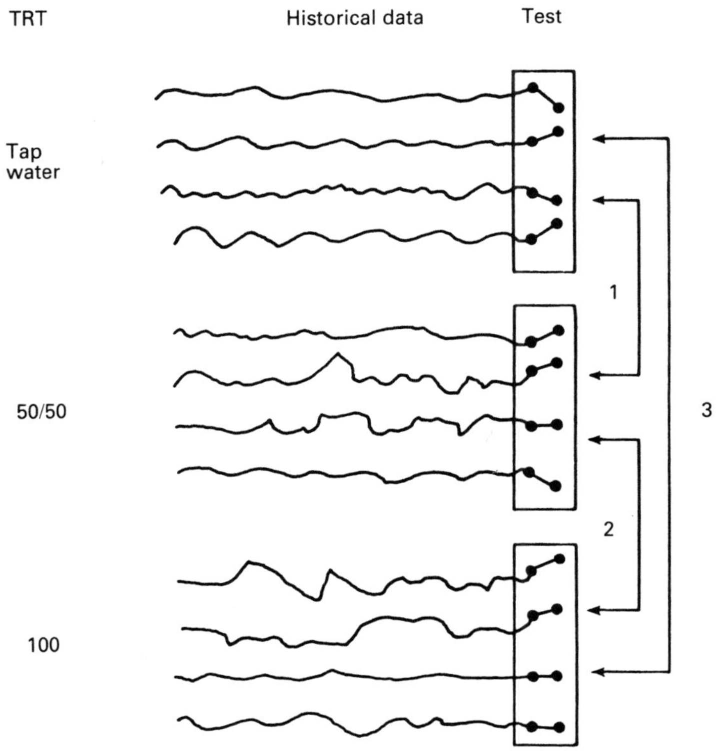

Fig. 4.5 — Diagrammatic representation of the 'delta' group-based method for evaluating when significant VF responses in the 50% and 100% effluent groups are occurring. Differences between the current and previous VF values for each fish within a group are used to evaluate intergroup differences. The three pairwise comparisons are noted with numerical values.

or a return to, group similarity. This is simply done by a quick check of the absolute values to see if group differences are indicated or not. If no differences are indicated, the delta response represents a shift back to group similarity, and is ignored.

4.3.3 Comparative responses

Application of analytical approaches to the March 9–14, 1985 data series suggested, most importantly, that the assumed control–treatment setting was not sound. This was indicated by the occurrences of what we consider invalid response patterns (the 'V' or the inverted 'V' response) when the absolute method was employed (Fig. 4.6). Violations of control assumptions, primarily that the tap water represents the quality of the effluent water environment less any system-related toxicity, detracts severely from the power of the group-based analytical procedures. Because the group-based approach represents a valuable family of analytical procedures for the continuous ABMS setting, modifications to the ABMS that would, among other things enhance soundness of the control-treatment setting, are being considered.

Nonetheless, certain anticipated characteristics of the various analytical procedures were apparent when these were applied to the March 9–14 data series when

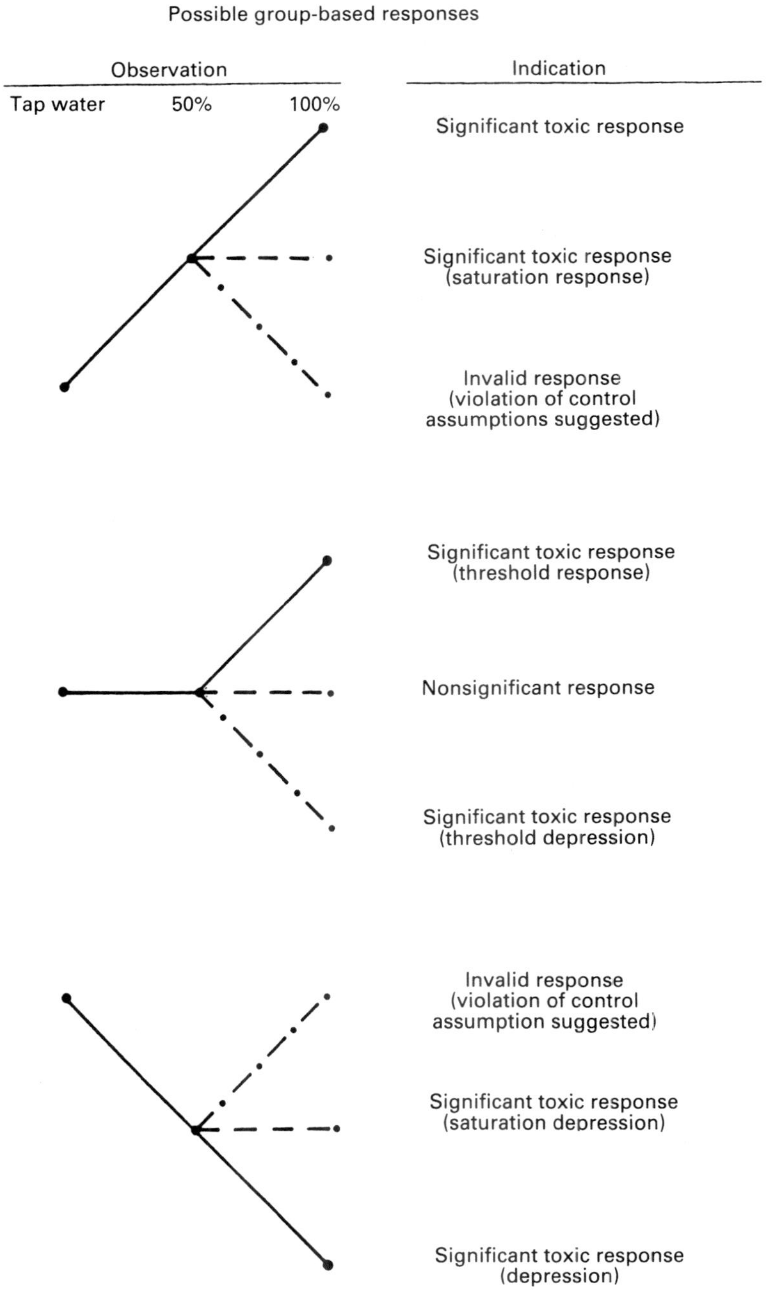

Fig. 4.6 — Representation of the nine possible response patterns that could occur among the three groups of bluegills with respect to the median VF value for each group on a given hour. An increasing or decreasing slope between two groups indicates the numerical direction of significantly different median VFs.

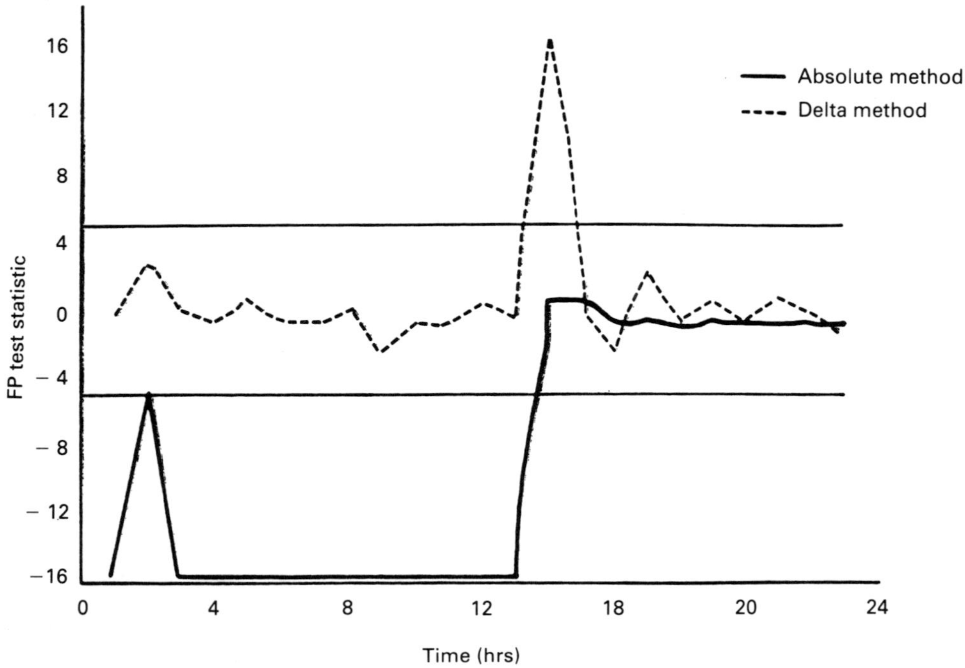

Fig. 4.7 — Contrast of the results of two group-based analytical procedures, the absolute and delta methods, applied to the same series of bluegill VFs (tap water versus 50% effluent group, March 10 data, Figs 4.1a,b). Portrayed are the Fligner–Policello (FP) test statistic values for each method by hour. Test values falling outside of the critical values (solid horizontal lines at ±4.86) indicate a significant difference between the median VF of the tap and 50% group for a particular hour. The conservative nature of the delta method relative to the absolute is emphasized. Only major shifts in the latter are detected by the delta method (note hour 13 on graph).

system toxicity was considered negligible (Bean, 1985) (Table 4.1). Using the criteria for establishing the significance of an alarm shown in Fig. 4.6, the absolute group-based procedure registered the greatest number of significant VF responses of the three procedures for the five-day period. For the absolute group-based procedure, a high incidence of invalid responses was observed throughout the first 24-hour segment followed by infrequent invalid responses over the remaining four days (18–24 and 5–94, respectively). The invalid response pattern showed the 50% effluent group to be significantly higher than the control and 100% groups, which were not significantly different from each other. Although the response was largely restricted to the first 24 hours, invalid control conditions undoubtedly pervaded the entire series to some extent, as indicated by the occasional appearance of the invalid pattern.

The delta method showed far fewer significant responses than its group-based counterpart, the absolute method, in the March 9–14 data series (Table 4.1). Significant responses were typically indicated by the delta method when the absolute responses shifted from nonsignificant to significant group differences, or when the

Table 4.1 — Summary of alarm responses of the three analytical procedures applied to VF data generated by the ABMS on March 9–14 and April 25–26, 1985

	Group-based procedures		Individual procedure
	Absolute	Delta	Time series analysis
March 9–14			
Number of intervals	144	145	138
Number of alarms	82	16	11
Percentage alarms	57	11	8
April 25–26			
Number of intervals	39	39	39
Number of alarms	11	7	17
Percentage alarms	28	18	44

form of group response patterns changed. Time series analysis (individual approach) which ignored the tap and 50% effluent groups, was unaffected by apparent problems with the control–treatment setting. During the initial 24 hours of the March 8–14 data series, time series analysis indicated that VFs for the bluegills in 100% effluent water did not exceed their 'normal' levels. From March 11, 0700h, only a single bluegill in 100% effluent water was exhibiting a VF beyond its 'normal' range. This low grade response indicated by time series analysis was registered as a significant VF response by the absolute group-based method. Again, problems in the control–treatment setting may have caused oversensitivity of the absolute group-based approach.

Consideration of the ABMS data series for April 25–26, 1985, where system toxicity was considered to be present, proved likewise to lead to new insights. A pronounced increase in VFs for fish in effluent water allowed an assessment of each procedure's ability to detect 'clear-cut' responses. All three procedures detected the sharp VF increase that occurred at 1600h on April 25, apparently in response to shifting pH (Fig. 4.2). A short-term recovery/episode was observed at 2100h on April 25, followed by another sharp increase and a second recovery. The delta group-based method identified all the crests and troughs of the response patterns, whereas the absolute group-based and the individual-based procedure (time series analysis) simply indicated continuous significant response conditions as indicated by an overall higher number of alarms relative to the absolute and delta group-based methods (Table 4.1). This evaluation demonstrates a greater sensitivity to important change in VFs by the delta group-based procedures. Following amelioration of present conditions that may be underlying the control–treatment problems in the ABMS system, relative performance of the three methods should be conducted on data where known toxic responses are, perhaps, less pronounced. Furthermore, the ability to track the performance of the various analytical methods from known nontoxic periods into periods where toxicity is present, would permit 'best-case' evaluations of selected analytical approaches.

An additional, and potentially very important, issue has emerged as a result of working with the April 25–26 data. The pronounced bimodal VF responses that were ovbserved for the 100% effluent fish were likewise observed for one of the four 50% effluent fish. This alone is not a curious result because conceivably one fish in the 50% group was more sensitive to a lower toxicant concentration than its three counterparts. A pronounced and temporally coincidental pattern, however, was also observed for one bluegill in the control group where no parallel toxicity should have been present. This observation suggests the possibility of 'stress transmittance' from one fish to another. Stressed fish often thrash with an intensity or frequency that is related to the undesirability of the environment. Such thrashing may well be perceived by nearby bluegills, not necessary through visual pathways. The possibility of stress transmittance in the ABMS system should be carefully considered and may, in part, be alleviated by the new exposure tank design which took into consideration isolation of the fish.

REFERENCES

Bean, D. J. (1985) *Environmental engineering services for Pine Bluff Arsenal, Arkansas, Task 001: Onsite services in support of Pine Bluff Arsenal's automated biomonitoring system.* Final report to Pine Bluff Arsenal, Pine Bluff, Arkansas. Battelle Columbus Division, Columbus, Ohio.

Cairns Jr, J., Thompson, K. W., Landers Jr, J. D., McKee, M. J. and Hendricks, A. C. (1980) Suitability of some fresh water and marine fishes for use with a minicomputer interfaced biological monitoring system. *Wat. Resourc. Bull.* **16**(3) 421–427.

Cairns Jr, J. and Garton, R. R. (1982) Use of fish ventilatory frequency to estimate chronically safe toxicant concentrations. *Trans. Am. Fish. Soc.* **111** 70–77.

Helwig, J. T. and Council, K. A. (eds) (1982) *Statistical Analysis System User's Guide*, 1982 edn. SAS Institute, Cary, North Carolina.

Hollander, M. and Wolfe, D. A. (1973) *Nonparametric Statistical Methods*. John Wiley, New York.

Thompson, K. W., Deaton, M. L., Foutz, R. V., Cairns Jr, J. and Hendricks, A. C. (1982) Application of time-series intervention analysis to fish ventilatory response data. *Can. J. Fish. Aquat. Sci.* **39** 518–521.

Part II

Automated biological systems using fish

5

Ventilatory and movement responses of rainbow trout exposed to 1,3,5-trinitrobenzene in an automated biomonitoring system

William H. van der Schalie and **Tommy R. Shedd**
US Army Biomedical Research and Development Laboratory, Fort Detrick, Frederick, Maryland 21701–5010, USA

Maurice G. Zeeman
US Food and Drug Administration, Environmental Impact Staff, CVM, HFV-152, Rockville, Maryland 20857, USA

5.1 INTRODUCTION

Immediate detection of developing toxic conditions in water can be critical in preventing fish kills resulting from effluent discharges or for avoiding the use of contaminated surface water in water treatment plants. A number of automated systems are presently in use that monitor the toxicity of water or wastewater by recording physiological or behavioral responses of fish. The most commonly monitored parameters include fish ventilatory rate, rheotactic response, and activity (van der Schalie, 1986).

Fish ventilatory rate is generally measured by placing electrodes in the water with the fish and electronically analyzing the electrical signal generated as the fish ventilate their gills. Although ventilatory rate is usually the only parameter monitored by this method, additional information can be obtained from the electrical signal on the strength of the ventilatory movements (ventilatory depth), the rate of coughing, and the presence of whole-body movement. This research was undertaken to determine: (1) which of these end points is most sensitive in detecting the toxicant 1,3,5-trinitrobenzene (TNB) in water, (2) what concentrations of TNB are detected by the biomonitoring technique relative to more commonly measured acute and chronic toxicity end points, and (3) how ventilatory and movement responses to TNB vary between two fish species (rainbow trout and bluegill) commonly used in automated biomonitoring systems. TNB was chosen because it is found in wastewaters resulting from the manufacturing process for 2,4,6-trintrotoluene (TNT), it is an environmental transformation product of TNT in aquatic systems (Burlinson, 1980), and there are data available on the toxicity of TNB to both rainbow trout and bluegills (van der Schalie, 1983; van der Schalie *et al.*, in press).

5.2 METHODS AND MATERIALS

5.2.1 TNB analysis

TNB was synthesized and recrystallized to a final purity of 99.97%. Stock solutions of TNB were prepared in deionized, distilled water. Test solution concentrations of TNB were analyzed by gas chromatography (Rosencrance and Brueggemann, 1986). The detection limit was 5 μg/l, and the average recovery rate for four spiked samples was 98.9% (±4.9%).

5.2.2 Water quality

Testing was conducted in well water having the following characteristics: hardness, 169 mg/l as $CaCO_3$; alkalinity, 220 mg/l as $CaCO_3$; pH 8.2; conductivity, 726 μSiemens/cm. Temperature was 12°C (±1°) and dissolved oxygen concentration ranged from 86% to 89% of saturation.

5.2.3 Rainbow trout

Rainbow trout (*Salmo gairdneri*) were obtained from the National Fisheries Center-Leetown, Kearneysville, West Virginia. Fish were held in well water for two months prior to testing. For the last two weeks of holding, trout were kept under the same constant light conditions used in testing to eliminate diurnal variations in ventilatory and movement patterns. Trout used in testing were approximately four months old, with an average weight and length of 4.1 g and 67 mm, respectively. Food consisted of pelleted trout food plus frozen brine shrimp prior to testing. Fish were not fed during testing.

5.2.4 Ventilatory test

Ventilatory test methods followed those described in van der Schalie *et al.* (in press). Ventilatory and movement patterns of trout were monitored using electrodes placed in the test chambers with the fish. Electrical signals generated by the fish were amplified, filtered to remove 60 Hz electrical interference, and passed to a Cromemco model Z-80 microcomputer that monitored ventilatory and movement information. The 28 test chambers, each housing one fish, were kept in plywood enclosures to minimize disturbances of the fish. Fluorescent bulbs provided constant light with an intensity of 500–1000 lux.

Test solutions were provided continuously by a proportional diluter at a rate of about 28 ml/min for each test chamber. There were five TNB concentrations, with a 50% dilution level between treatments, plus controls. Five trout were tested at each treatment level except at the highest TNB concentration, where three trout were exposed. TNB samples were taken for analysis from each concentration tested on the first two days of exposure.

Fish were placed in the test chambers and allowed to acclimate to the dilution water for three days prior to the start of ventilatory signal monitoring. All fish were then monitored for a four-day pre-exposure period to establish normal baseline ventilatory values for each fish, followed by a three-day TNB exposure period. Control fish were held in dilution water throughout the test.

Fish ventilatory signals were monitored continuously by the microcomputer. The ventilatory and movement paramaters were calculated every 15 seconds, and the summed 15-second data were recorded every 15 minutes. The number of ventilatory and cough peaks per minute were determined, as was the average height of the ventilatory peaks (over 15 minutes) and the percentage of the time that each fish was moving (number of 15-second movement periods divided by 60 (total periods in 15 munutes) multiplied by 100). Ventilatory parameters were not determined for any 15-second segment in which movement was detected. The 15-minute data records were stored on magnetic tape, then passed to another computer (AMDAHL 470-V7) for statistical analysis.

Ventilatory data were smoothed using a moving average technique. Individual 15-minute data points were replaced with the average value over the previous four hours (16-data points). With each new 15-minute data point, the first data point was dropped and a new mean value computed. The moving average process was restarted at the beginning of the exposure period so that no pre-exposure data would be included in the analysis of the exposure period.

After the moving average technique was applied to the data, maximum and minimum values for each ventilatory parameter and fish were determined for both exposure periods. Since toxicant-induced effects on ventilatory or movement patterns were frequently manifested as large initial changes followed by a return toward baseline levels, comparing changes in maxima and minima was more appropriate than comparing differences between the whole pre-exposure and exposure periods. Changes in ventilatory patterns were determined by subtracting exposure minima from pre-exposure minima and exposure maxima from pre-exposure maxima for each fish, then testing for significant changes in each TNB-treated group relative to the controls using the General Linear Models (GLM) procedure of the Statistical Analysis System (SAS Institute, 1985). Analysis of variance techniques were used to determine overall treatment effects (p=0.05) using either linear or quadratic models of the $\log_{10}$ of the TNB concentration and the ventilatory or movement parameters. Pairwise comparisons between each treatment and the control were made using Student's t-tests. The significance level for the pairwise comparisons was set at p=0.01 by application of Bonferroni's correction for simultaneous comparisons.

5.3 RESULTS AND DISCUSSION

A summary of the data from the four days prior to TNB exposure is provided in Table 5.1. The wide ranges of movement, cough, and, to some extent, ventilatory rates are related to the high variability in the electrical signals generated by the trout. Although there was substantial variation among fish, overall trout displayed greater whole body movement (16.6%) than was found for bluegills (<1%) during pre-exposure periods in previous tests (van der Schalie, 1980). These high movement rates can also carry over into increased cough rates, since movement peaks that occur

Table 5.1 — Ventilatory and movement parameters during the pre-exposure period

Parameter error	Mean[a]	±Standard Error	Range
Ventilatory rate (per min)	62.2	4.60	28.0–127.6
Ventilatory depth (volts)	1.41	0.11	0.84–3.38
Cough rate[b] (per min)	0.85	0.11	0.23–2.38
Movement rate (%)	16.6	2.64	2.2–41.6

[a]N=28 fish.
[b]Data from 27 fish are summarized; one fish had a cough rate of 42.3/min (see text).

in only a small portion of the 15-second blocks of data monitored by the computer may mistakenly be counted as coughs. The one fish that had the unreasonably high cough rate (42.3/min) also had one of the higher movement rates (24.9%). In spite of this potential interaction, cough rate primarily reflects cough-like peaks in the electrical signal, whereas movement rate primarily shows the extent of distortions of the electrical signal caused by whole body movements.

Ventilatory and movement parameters for the control fish tended to decrease in the exposure period relative to the pre-exposure period (Tables 5.2 and 5.3). This

Table 5.2 — Differences between pre-exposure and exposure ventilatory and movement parameter minima

	Mean difference between minima[a]			
TNB Conc. (mg/l)	Ventilatory rate	Ventilatory depth	Cough rate	Movement rate
0.567	6.82 (39.59)	0.95 (4.65)	−0.02 (0.31)	14.97[b] (14.89)
0.272	16.75 (13.84)	2.30 (5.03)	0.18 (0.41)	3.13 (8.89)
0.137	−1.10 (11.55)	2.83 (4.94)	0.21 (0.75)	3.19 (7.61)
0.058	−7.06 (4.79)	0.66 (1.57)	−2.72 (6.01)	−3.29 (3.78)
0.034	−14.06 (11.16)	−0.56 (0.75)	−0.06 (0.17)	−3.90 (7.18)
<0.005[c]	−1.81 (12.09)	−1.04 (0.72)	−0.10 (0.13)	−1.48 (2.93)

[a]Values represent the mean ±(standard deviation) for five fish (three at 0.567 mg/l) of: (minimum value during the exposure period) minus (minimum value during the pre-exposure period).
[b]Treatment was significantly different from the control (p=0.01), using Student's t-test with Bonferroni's correction for simultaneous comparisons.
[c]Control treatment.

could be because of continuing acclimation to the test chambers, or because the fish were not fed during the test, or both. A similar pattern was also noted for control bluegills in an earlier TNB test (van der Schalie *et al.*, in press).

In contrast to the control fish, when TNB-exposed trout (or bluegills) exhibited a

Table 5.3 — Differences between pre-exposure and exposure ventilatory and movement parameter minima

TNB conc. (mg/l)	Mean difference between maxima[a] Ventilatory rate	Ventilatory depth	Cough rate	Movement rate
0.567	38.89[b] (17.92)	8.73 (18.89)	3.09 (5.90)	34.18[b] (10.81)
0.272	20.65[b] (32.97)	35.45 (34.68)	7.84 (3.83)	33.82[b] (16.52)
0.137	9.41 (18.21)	13.14 (26.30)	2.54 (3.16)	12.82[b] (22.63)
0.058	−9.41 (24.25)	2.00 (5.60)	−1.83 (2.40)	1.14 (11.13)
0.034	1.11 (27.70)	−1.07 (5.14)	2.05 (5.51)	−9.33 (5.42)
<0.005[c]	−11.53 (9.33)	−6.50 (8.27)	−0.46 (1.23)	−10.34 (8.13)

[a]Values represent the mean ±(standard deviation) for five fish (three at 0.567 mg/l) of: (minimum value during the exposure period) minus (minimum value during the pre-exposure period).
[b]Treatment was significantly different from the control (p=0.01), using Student's t-test with Bonferroni's correction for simultaneous comparisons.
[c]Control treatment.

significant response to TNB, it was always as increased rates of ventilatory or movement parameters. Trout ventilatory and movement rate minima increased significantly with increasing TNB concentrations in spite of high variability between fish (Table 5.2). (With TNB-exposed bluegills, no significant increases in minima were found (van der Schalie *et al.*, in press).) Although the overall ventilation-rate minima model was significant (p=0.02, R-square=0.26), no significant differences were found in pairwise comparisons between controls and TNB-exposed fish. Movement rate minima were significantly increased at the top concentration, and the overall model was highly significant (p=0.01, R-square=0.37). Although there was an apparent trend towards increased ventilatory depth minima in the TNB-exposed fish, it was not statistically significant.

Maxima values increased significantly in the TNB-exposed fish (Table 5.3) for ventilatory rate (p=0.02, R-square=0.26), ventilatory depth (p=0.04, R-square=0.28), cough rate (p=0.04, R-square=0.22), and movement rate (p=0.0001, R-square=0.65). As with the minima, variability in maxima changes between fish at each TNB concentration was high, so R-square values were generally low. The strongest TNB effect was on movement rate; fish at the top three concentrations significantly increased their movement rates during TNB exposure, whereas movement rates in the controls actually declined. Ventilatory rate was increased significantly only at the top two TNB concentrations. As with the ventilatory depth minima, ventilatory depth maxima increased with increasing TNB concentrations except at the top concentration, but the changes in ventilatory depth maxima were statistically non-significant.

The overall sensitivity of rainbow trout to TNB was the same as of bluegills; the range between the lowest concentrations causing a significant effect on any ventilatory or movement paramater and the highest concentration causing no effects was 0.06–0.13 mg/l for bluegills and 0.06–0.14 mg/l for trout (Table 5.4). Although the

Table 5.4 — Toxicity of TNB to fish

Test type	Fish	Length of exposure (days)	Test end point	Results[a]	Reference
Static acute	Bluegill	4	LC_{50}	0.85 (0.52–1.38)	van der Schalie (1983)
Static acute	Rainbow trout	4	LC_{50}	0.52 (0.32–0.80)	van der Schalie (1983)
Early life stage	Rainbow trout	71	NOEL[b]–LOEL	0.08–0.17	van der Schalie (1983)
Early life-stage	Fathead minnow	32	NOEL–LOEL	0.08—0.12	van der Schalie (1983)
Ventilatory	Rainbow trout	3	NOEL–LOEL	0.06–0.14	This report
Ventilatory	Bluegill	6	NOEL–LOEL	0.06–0.13	van der Schalie *et al.* (in press)

[a]TNB concentration in mg/l (95% confidence limits, if applicable).
[b]NOEL — no observed effect level (highest concentration tested causing no statistically significant differences from the controls);
LOEL — lowest observed effect level (lowest concentration tested causing statistically significant differences from the controls).

bluegills were exposed for a longer period of time than the trout (six days as compared with three days for trout), all significant responses by the bluegills were evident within the first three days after the start of TNB exposure (van der Schalie *et al.*, in press).

When the sensitivity of fish ventilatory and movement responses to TNB is compared to more traditional toxicity measurements (Table 5.4), the no effect–effect range for the ventilatory and movement responses is found to be virtually the same as that for TNB chronic toxicity estimated from trout and fathead minnow (*Pimephales promelas*) early-life-stage tests. Not all biological monitoring systems employing fish ventilatory and movement sensors have been able to detect toxicants at chronically toxic levels, but nearly all of these systems rely on only one response end point (van der Schalie, 1986). A comparison of the bluegill and trout TNB tests clearly shows the importance of relying on more than one end point in evaluating toxicant responses. With bluegills, ventilatory depth was the most sensitive indicator of TNB effects, but changes in cough and movement rates were found only at the top concentration tested (0.61 mg/l), and there was no effect on ventilatory rate. In contrast, movement rate was most sensitive for trout, followed by ventilatory rate, with no changes detected in either ventilatory depth or cough rate. With some chemicals, however, cough rate has been found to be an extremely sensitive indicator of the presence of toxic materials in water (Drummond and Carlson, 1977).

Even though the use of multiple monitoring parameters may increase the sensitivity of biological monitoring systems, and fish systems using multiple parameters detected TNB at chronic toxic concentrations, the biological significance of these ventilatory and movement changes is uncertain. Graphs of ventilatory and movement parameters versus time during TNB exposure show that many of the responses of both trout and bluegill are transient (van der Schalie, unpublished data); they peak and then begin to return toward pre-exposure levels. Similar transient changes were observed for cough rates in fish exposed to certain toxicants

(Drummond and Carlson, 1977). Whatever the long-term meaning of fish ventilatory and movement changes, however, it is clear that automated biomonitoring systems are a unique means of continuously monitoring the toxicity of water and wastewater and may be capable of predicting the chronic effect levels for some toxic chemicals. Use of multiple parameters can increase the sensitivity of the monitor and improve the chances of detecting the presence of unsuspected toxic materials in water.

DISCLAIMER

'The opinions or assertions contained herein are the private views of the authors and are not to be construed as official or as reflecting the views of the Department of the Army or the Department of Defense.'

'In conducting the research described in this report, the investigators adhered to the *Guide for the Care and Use of Laboratory Animals* as promulgated by the Committee on the Guide for Laboratory Animal Facilities and Care of the Institute of Laboratory Animal Resources, National Research Council.'

REFERENCES

Burlinson, N. E. (1980) *Fate of TNT in an aquatic environment.* Report NSWC TR 79-445. Naval Surface Weapons Center, Silver Spring, MD.

Drummond, R. A. and Carlson, R. W. (1977) *Procedures for measuring the cough (gill purge) rates of fish.* EPA-600/3-77-133. US Environmental Protection Agency, Washington DC.

Rosencrance, A. B. and Brueggemann, E. E. (1986) *Liquid and gas chromatographic determinations of 1,3,5-trinitrobenzene in water.* Technical report 8601. U.S. Army Biomedical Research and Development Laboratory, Fort Detrick, Frederick, MD.

SAS Institute (1985) *SAS users guide: statistics, version 5.* SAS Institute, Cary, North Carolina.

van der Schalie, W. H. (1980) A new technique for automatic monitoring of fish ventilatory patterns and its possible use in screening tests for chronic toxicity. In J. G. Eaton, P. R. Parrish, and A. C. Hendricks (eds) *Aquatic toxicology,* ASTM STP 707, pp. 233–242. American Society for Testing and Materials, Philadelphia.

van der Schalie, W. H. (1983) *The acute and chronic toxicity of 3,5-dinitroaniline, 1,3-dinitrobenzene, and 1,3,5-trinitrobenzene to freshwater aquatic organisms.* Technical report 8305, AD A138408. US Army Medical Bioengineering Research and Development Laboratory, Fort Detrick, Frederick, MD.

van der Schalie, W. H. (1986). Can biological monitoring early warning systems be useful in detecting toxic materials in water? In T. M. Poston and R. Purdy (eds) *Aquatic toxicology and environmental fate: ninth volume,* ASTM STP 921, pp. 107–121. American Society for Testing and Materials, Philadelphia.

van der Schalie, W. H., Shedd, T. R. and Zeeman, M. G. (in press) Ventilatory and movement responses of bluegills exposed to 1,3,5-trinitrobenzene. In W.

Adams, G. Chapman, and W. G. Landis, (eds) *Aquatic toxicology and hazard assessment, tenth volume,* ASTM STP 971, American Society for Testing and Materials, Philadelphia.

6

The WRc Fish Monitor and other biomonitoring methods

G. P. Evans,
Water Research Centre, Henley Road, Medmenham, PO Box 16, Marlow, Buckinghamshire. SL7 2HD. United Kingdom
J. F. Wallwork,
Sunderland and South Shields Water Company, 29 John Street, Sunderland. SR1 1JT. United Kingdom

6.1 INTRODUCTION

The water sources used for drinking water supply which are most likely to suffer the effects of pollution or major natural changes of quality are almost invariably lowland rivers. One-third of the treated water supplied in the United Kingdom is derived from approximately 140 lowland river intakes of which about one-half have some form of raw water storage to act as a buffer between the river and the associated treatment works. During 1983 about 12 000 notified pollution incidents occurred in the rivers controlled by the 10 English and Welsh Regional Water Authorities, compared with around 6000 during the period April 1977 to March 1978 when approximately 100 intake closures occurred of which 80% were caused by pollution incidents. The pollutants involved in the closures are summarized in Table 6.1.

Table 6.1 — Pollutants involved in intake closures

Pollutant	Percentage occurrence
Fuel oils	40
Farm waste	10
Sewage	10
Plating effluent	5
Phenols	5
Other chemicals from industry or road accident	30

Continuous protection of river intakes against a very varied and growing range of pollutants is crucial to the maintenance of final water quality in terms of its safety and acceptability to the consumer. Pollution following road or rail accidents, failures of chemical or agricultural storage vessels, activities of the emergency fire services or arising from wilful damage all indicate an increasing need for the continuous, online monitoring of lowland river raw water quality as a means of detecting pollution. Natural variations in river water quality may also be sufficiently large or frequent to act as an important secondary phenomenon for which online monitoring would be an operationally valuable facility.

Fish monitors (Morgan, 1975; Miller, 1977; Van der Schalie *et al.*, 1979; Van Hoof, 1980; Gruber and Cairns, 1981a,b; Thompson *et al.*, 1982; Cairns and Garton, 1982; Van der Putte *et al.*, 1982; Evans and Johnson, 1984, 1985; Wallwork *et al.*, 1977) have the advantages of using a broad spectrum sensor which is at least as sensitive as human beings to a wide range of substances and which can be incorporated into a system which is reliable, rapid in response and capable of full automation for warning purposes. However, there are a number of points which must be borne in mind when choosing an appropriate system.

6.2 CHOICE OF MONITORING SYSTEM

6.2.1 Fish species

This should be capable of a healthy existence in the range of ambient temperatures that characterizes the river concerned and for this reason it should ideally be indigenous. Full behavioral information should be available and fish should be easily obtainable from a genetically consistent stock which is not prone to disease.

6.2.2 Physiological parameter

The WRc Mk III Monitor (Evans and Johnson, 1984, 1985) uses ventilation frequency because it is the most specific and sensitive indicator of fish condition and also because it is easily measured, i.e. it provides a strong electrical signal in comparison with that derivable from the heart beat.

6.2.3 Method of comparison with normal behavior

Absolute limits have the appeal of simplicity and can be derived from the literature and then related to a range of electrical signal strengths. However, the natural variation between individual fish makes it difficult to set generally applicable limits.

6.2.4 Use or otherwise of a control water stream

A control stream against which the behavior of the monitor fish can be evaluated could, for example, consist of water from the associated treatment works or other local mains water supply. Whilst this has the advantage that the control fish are used in drinking water of quality characteristics identical to those of the supply to consumers, it is not a true control in that control stream quality should duplicate that of the monitor stream, except for polluting material

A possible solution might involve a stream of raw water derived from a holding reservoir, i.e. delayed for a preset time period so that the control stream quality would necessarily represent conditions prior to a pollution incident occurring. Extra complication and expense would, however, be required because of the need for an automatic stream changeover system for any fish physiological parameter other than that of the physical avoidance of incoming pollution. There must also be a sufficient flow for this purpose which may therefore further increase the cost.

The above difficulties are overcome in the WRc Mk III Monitor by using the comparison of present with historical behavior over an arbitrary time period on an individual fish by fish basis. The procedure is therefore unaffected by variations of different individuals' characteristics from an arbitrary standard but may be affected by natural phenomena such as the diurnal cycle which may cause variations of frequency with time on a regular though predictable basis.

6.3 DESCRIPTION OF THE MK III MONITOR

6.3.1 Background

Before describing the Mk III Fish Monitor, it is instructive to give an outline account of Mks I and II which provided the background of research and operational information and experience on which the Mk III is based

The Mk I monitor (Miller, 1977) was based on the measurement of the total physiological activity of each of six rainbow trout (*Salmo gairdneri*) placed in individual tanks as two physically separated groups of three fish each. The tanks contained a pair of horizontally mounted electrodes at the top and bottom of each tank; signals from these electrodes were amplified and the amplitude compared by analog means with limits that had been preset on potentiometers. Four signal strength bands were defined, namely Normal 1, Normal 2, Hyperactivity and Death. The River Wear monitoring station in North-East England (Wallwork *et al.*, 1977; Wallwork, 1980; Wallwork and Ellison, 1983) used a numerical scoring system as the operational means of assessing an alarm condition: a source of 8 for Hyperactivity, 4 for Death and 1 each for the Normal 1 and 2 states was allocated to each fish. If the total equalled or exceeded 23, the works intake would be closed. No pollution incident requiring intake closure has been experienced, but the works has been taken out of supply, with diversion of partially treated water to river, for scores of less than 23.

However, the system had several disadvantages. False alarms occurred fairly regularly as a result of moderately severe natural variations in water quality, e.g. turbidity and temperature, and below about 4°C fish activity sharply decreased and became electronically almost indistinguishable from death (the 'sleep' state). Setting the potentiometers was problematic in that the settings had to be altered according to whether or not winter temperatures prevailed. The monitor lacked sensitivity and discrimination: some minor pollution incidents from different chemicals are known not to have been detected.

The Mk II version was an improvement in that it was designed to measure ventilation and cardiac frequency, though by analog means. The hydraulics of the system were also improved but general development was halted following a number of major improvements in design concepts and data acquisition and analysis, with the result that work began shortly afterwards on Mk III (Evans and Johnson, 1984, 1985).

6.3.2 Design

Extensive redesign of tanks and electrodes has resulted in the use (as before) of two groups of tanks, but with four fish in each group. Digital processing of fish signals allows the use of simpler amplifiers and the introduction of silver/silver chloride electrodes gives predictable and small d.c. voltage offsets at the entry to the amplifiers. The amplifiers are a.c. coupled to ensure that 'limiting' (i.e. reaching an output voltage equal to either the +ve or −ve power rail voltages) is less likely and the amplifier gain has been reduced to 10 000 to keep the signal within limits except during the most vigorous fish movement. Signal strength has an amplitude of 10–50 μV and the amplified signals are converted to digital form with an eight-channel eight-bit A to D converter which gives a digital output from −128 to +127.

6.3.3 Determination of ventilation (breathing) frequency

A fast Fourier transform is used to obtain the power spectral density (PSD) as a function of frequency; this is known as the power spectrum. The power spectrum is scanned over a credible fish breathing fequency range, e.g. 0.3–3.0 Hz, and if a maximum is found in that 'window' which is greater than the preset threshold in the software, the frequency is deemed to be calculable. If there is no such maximum, then the frequency is set to zero. If there is such a local maximum, the 'centre of gravity' (in terms of frequency) of a range of the spectrum is evaluated and is deemed to be the fish breathing frequency.

The total power (which is similar to the fish 'activity' in the Mk I monitor) is also calculated as the integral of the PSD over the credible fish breathing frequency range.

The fish 'ratio' per individual fish is then calculated as power at breathing frequency divided by total power. This is typically 0.4, but if the frequency is not calculable because the maximum power is too low then it is set to zero.

6.3.4 Manner of use of ventilation frequency, power and power ratio

6.3.4.1 Ventilation frequency

This is the most sensitive parameter of pollution detection and is used in a percentile test which detects statistically significant changes in frequencies via an instrumental comparison with historical data. Frequencies are logged every minute into a 120-minute-long rolling store, the oldest 60 frequencies (the background set) are inspected at minute intervals and the 5th and 95th percentiles calculated. The 10 most recent frequencies (the inspection set) are then examined to determine how many are outside the percentile limits. Each fish thus has a 'score' between 0 and 10 and the monitor as a whole between 0 and 80. A binomial series predicts how many excursions, i.e. consecutive exceedings of the warning threshold, may be expected by chance. A change score of 22 or more/80 is vanishingly unlikely and corresponds to a false alarm of one per year, assuming that the test is performed once per minute. An alarm is raised if (say) 10 consecutive tests exceed the 22/80 limit, the precise number of such tests being user-configurable.

6.3.4.2 Power

Power is normally used for the purpose of detecting the death of all eight fish. The high and low limits are absolute and user-configurable, the low limit being set to indicate death; although the high limit can be used to indicate hyperactivity, frequency is a better choice of parameter for this purpose.

6.3.4.3 *Ratio*

The use of power ratio is essentially for fish husbandry, i.e. to detect individual sick or dead fish. It is usually configured for each fish separately and serves as an early warning of the need to exclude one or more fish from the statistical analysis and therefore to consider exchanging the fish concerned for fresh stock (see Section 6.4.3).

6.4 THE INTAKE PROTECTION SYSTEM (IPS)

The Mk III monitor is incorporated into a comprehensive IPS having the following features:

(1) Up to 16 other water quality instruments can be simultaneously monitored, e.g. instrumentation for dissolved oxygen (DO), Temperature, pH, conductivity or turbidity.
(2) Readings of fish parameters (frequency, power and ratio) and from other water quality instruments are made every minute. In addition, five-minute averages are calculated and logged every five minutes and the results are incorporated in a 24-hour rolling store of data holding the 288 most recent five-minute values.
(3) The frequency test can be carried out on the five-minute as well as on the one-minute data which is useful for detecting longer-term trends.
(4) Hardware consists of a 256K micro PDP 11/23 microcomputer, console local terminal, printer and remote terminal connected via a modem link.
(5) The software has been written by WRc Environment in FORTRAN and Macro-11 for the 11/23 processor operating under the RT-11 operating system.
(6) The system is menu driven through single-letter commands.

6.4.1 Effect of an alarm

An audiovisual alarm is given at the console and the remote terminal with simultaneous logging of data on the printer. Contact closures are also available for automatic sampling of river water, autodial telephone, telecommunications, a warning bell, etc.

6.4.2 Data analysis and display

By use of the menu-driven command interface the operator can plot graphs on the visual display unit of any parameter over a one-, two-, six- or 24-hour time period. Tables of numerical data are viewable from the present time backwards over a two-hour period at a one-minute resolution or over 24 hours for five-minute averages.

6.4.3 User configuration and control options

The results from individual instruments can be registered as invalid and thereby excluded from the alarm-raising process. The invalidity of the instrument can be deemed to be permanent or for a period of chosen duration, e.g. during the backflushing of sample lines.

From time to time the signal from individual fish may be too weak for the calculation of frequency and for those fish on those occasions a frequency of zero will be set in the rolling store. The zero values are ignored in the statistical test on the frequencies which compares the inspection sets with the background sets. However, a persistently weak fish will trigger a ratio alarm for that individual fish. The operator can register the signal as invalid and thereby permanently exclude it from consideration until a convenient time for replacement of that fish.

Alternatively, the data may be logged but the alarm checking of that particular variable can be suppressed, which as the advantage that processes such as fish acclimation can be observed even though they may not give rise to alarms.

Both data invalidation and the suppression of alarms can be introduced with 'timeout', i.e. for a chosen time period, and a special aspect of this feature is the provision of a single command for use when test fish are exchanged with stock fish. In these circumstances, incoming stock fish might be expected to be disturbed, and in any case it is necessary to build up an historic base for the new fish. The user designates as invalid one of the two cubicles whose group of four fish is to be exchanged and data from the new fish in that cubicle are not used as a contribution to the alarm statistics for a designated time period, e.g. 12 hours. During this period the monitor is a four-fish monitor.

6.5 POLLUTANT TESTING PROCEDURES

WRc has carried out a series of tests (Evans *et al.*, 1986) designed to determine the response of the monitor to various potential pollutants (see Table 6.2) and the following is a summary of the procedures which have been used in pollutant testing.

Fish are held in stock tanks supplied with a continuous flow of dechlorinated tap water. The tanks are aerated to guard against low oxygen concentrations developing in the event of a failure of the water supply. The fish are fed with commercial trout pellets twice weekly and accumulated debris is removed prior to feeding.

The fish to be used in the evaluation are transferred to the individual tanks and allowed to acclimatize over a three-day period. Their ventilation frequencies over this period are monitored and fish with particularly weak or variable signals are replaced.

During the following two weeks pollutant tests are carried out on alternate days (excluding weekends); this permits evaluation at six concentrations of pollutant. The fish are rested for a day between tests during which time their ventilation frequency usually returns to 'normal'. The fish are not fed at any time during the two-week period. They are returned to a separate stock tank following the experiments, and observed for any adverse long-term effect which could be attributed to the pollutant. The fish are then humanely destroyed in a solution of ethyl 4-aminobenzoate (Benzocaine).

Each experiment consists of a four-hour period which is divided into three parts. First a 90-minute period of normal conditions establishes a stable frequency background period with which to compare the effect of the pollutant; then the pollutant is dosed for one hour at a nominally constant concentration; finally a 90-minute period with no dosing ensues so that the recovery of the fish may be observed. In the control periods on either side of the four-hour experiment the frequency is measured at five-minute intervals unless the fish are showing signs of distress, in which case the measurements are continued at one-minute intervals.

In a series of experiments to be carried out at different concentrations, the doses usually start at the lowest concentration and are increased over the two-week period to the maximum. All pollutant tests are carried out in dechlorinated tap water to prevent the possibility of a response due to substances other than the dosed pollutant. The temperature, DO concentration, pH and conductivity of the water are measured at five-minute intervals during experimental and control periods and these values are logged together with the ventilation frequencies.

6.5.1 Results

Fig. 1 is an example of a pollutant test. Gamma-HCH (hexachlorohexane) is extensively used in the UK as an insecticide in a wide range of formulations and is highly toxic at low concentrations to insects and many other forms of life. The gamma-HCH used was contained in the ICI formulation 'Gamma-Col' which contains 800 g gamma-HCH/l together with a range of unspecified compounds added to give the required wetting activity when used as an insecticide. The stock dose solution was made up in distilled water and stored in a cool, dark place until required. The 96-hour LC_{50} of gamma-HCH for various freshwater fish species covers the range 128–2450 μg/l and the concentration used in the present example was 60 μg/l.

Subsequent analysis of samples collected during the test showed that the concentration of gamma-HCH rose to 60 μg/l within about 10 minutes of the start of the dose period and fell to zero within about 10 minutes of its end.

The upper part of the figure shows plots of the ventilation frequencies estimated at one-minute intervals for all the fish except fish 4 (six fish tanks were used in the testing system) for which signal processing was temporarily unavailable. Pre-dose frequencies vary between fish but in general remain constant.

Reaction to the dose begins 10 to 15 minutes after the start of dosing and takes the form of either increased variability, or a trend, of both of these. The presence of the pollutant has clearly affected the ventilation frequencies of the fish and similar responses have been detected (see Table 6.2), for other pollutants.

A common use of fish in a water treatment plant has been to keep a number of fish in a tank in the control room. Some of the water under treatment passes through the tank and the fish are observed for abnormal behavior or death. It is possible that, in the event of fish death, contaminated water has been present for some time and that it has already passed into the distribution system and so the level of pollutant detected will be at best a short term LC_{50} value. The sensitivity of the Mk III Fish Monitor may be assessed by comparing its detection limits with typical LC_{50} values and the two sets of figures are shown in Table 6.2.

As can be seen from the table, in all cases the Mk III Fish Monitor is considerably more sensitive than LC_{50} tests. As one would expect, detection levels vary considerably from one pollutant to another and it may be that for some compounds the detection limit is higher than levels at which there is cause for concern. In such cases

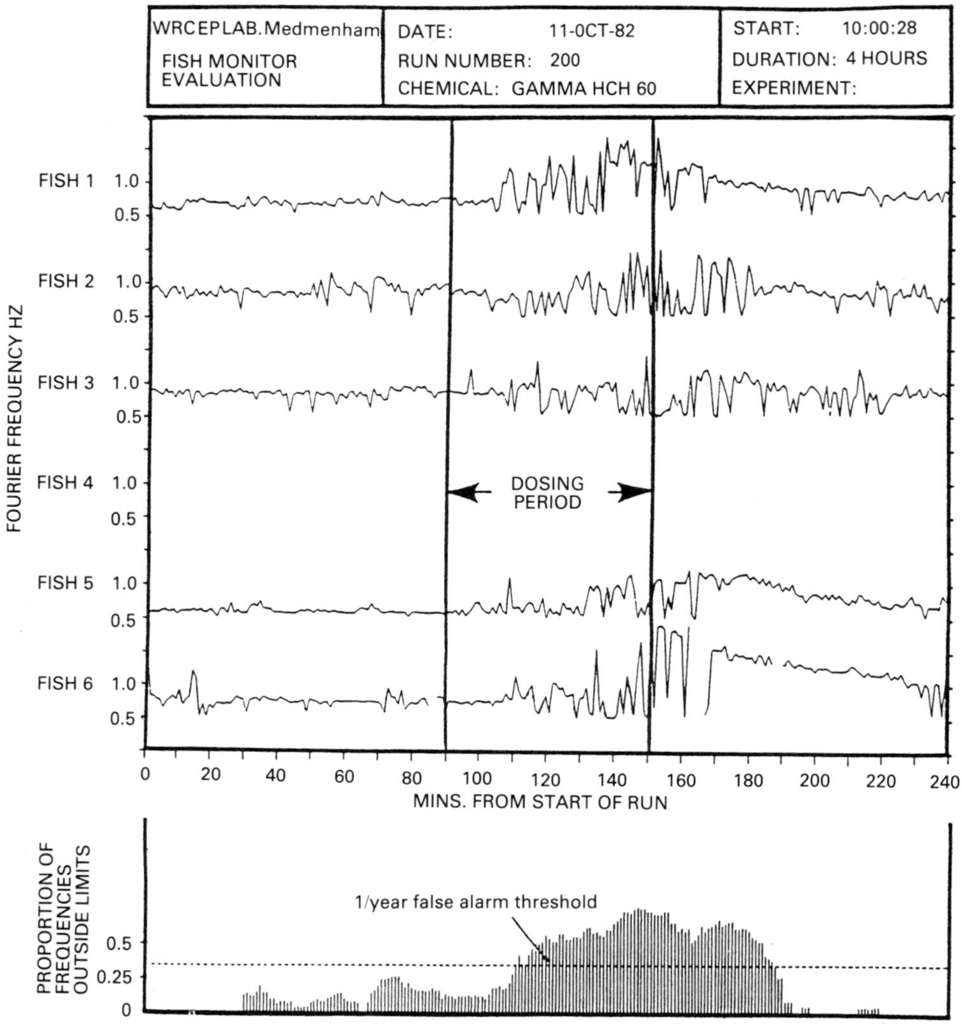

Fig. 6.1 — Four-hour pollutant test run: 60 μg/l γHCH closed for 1 hour.

the Fish Monitor will not be a suitable detection tool and another method of detection must be sought.

6.6 OPERATIONAL EXPERIENCE

6.6.1 Field trials

To assess the reliability and performance of the monitor under realistic conditions and to demonstrate an integrated protection system, a series of field trials was carried out over a period of 18 months at six treatment works intakes in different areas of

Table 6.2 — Comparison of typical LC_{50} values with Mk III Fish Monitor response times (Evans *et al.*, 1986)

	LC_{50}		WRc Mk III	
	Concentration (mg/l)	Time (hours)	Concentration (mg/l)	Time (mins)
Ammonia	0.16–0.49	96	0.11	17
Phenol	9.3	48	2	28
Paraquat	32	96	0.8	26
2,4,6-Trichlorophenol	10.0	24	0.2	20
γ-HCH	0.06	96	0.003	21
Pentachlorophenol	0.25	48	0.14	33
Formaldehyde	610	96	100	38
Diesel oil	—	—	11[a]	35
Petrol	—	—	3[a]	60

[a] μl/l.

England. During the field evaluation, a number of modifications were made to the system software and hardware and since completion of the evaluation the commercially available IPS has been developed at WRc Environment.

6.6.2 User experience

In the UK, seven systems have been installed during 1985–1986 and a further five were on order as at 30 September 1986 (see Fig. 6.2).

The first system to be installed was at the intake of a treatment works of the North-West Water Authority on the River Dee in North Wales following a major pollution incident in January 1984 (North West Water/Welsh Water, 1984; Water Authorities Association/Water Companies Association, 1984; Rushbrooke and Beaumont, 1986). At that time, an undetected phenol spillage occurred at a concentration of about 1 mg/l in the river and passed through the treatment scheme at the works. The resulting chlorphenols were present in the 320 Ml/day supply to the city of Liverpool at a concentration of up to 0.5 mg/l.

6.6.2.1 Alarms of known cause

Fish frequency alarms have occurred as a result of sudden changes of river water quality and flow largely associated with recent rainfall, e.g.

(1) An increase in ammonia content (cow slurry following heavy rainfall and a thaw) from less than 0.1 mg/l to 0.45 mg/l together with a turbidity increase from 15 NTU (Nephelometric turbidity units) to 100 NTU. The response of the monitor was almost immediate (20 min).

(2) A turbidity increase from 10 to 750 NTU during four hours. Alarms which started at the beginning of this period lasted for half an hour.

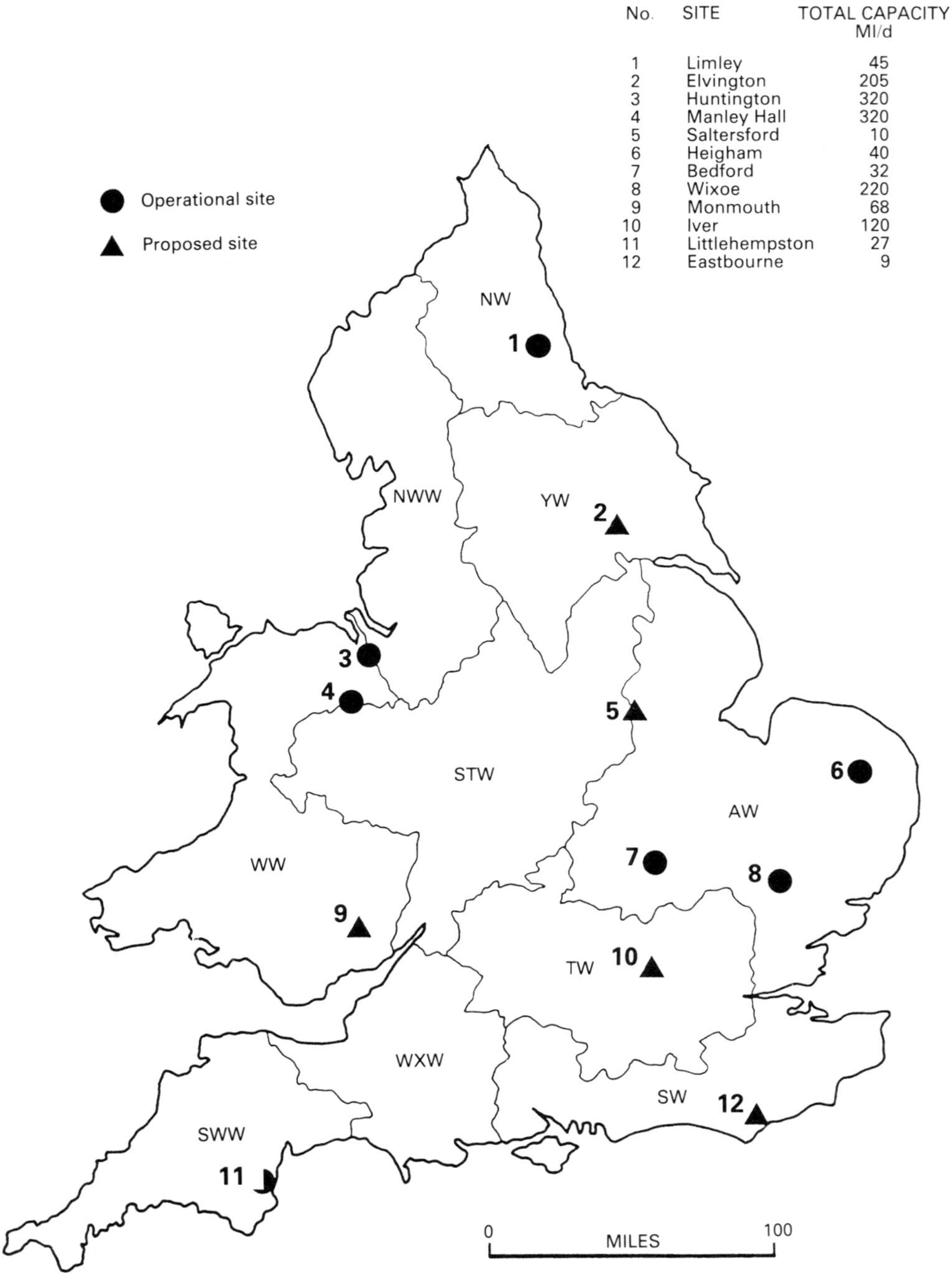

No.	SITE	TOTAL CAPACITY Ml/d
1	Limley	45
2	Elvington	205
3	Huntington	320
4	Manley Hall	320
5	Saltersford	10
6	Heigham	40
7	Bedford	32
8	Wixoe	220
9	Monmouth	68
10	Iver	120
11	Littlehempston	27
12	Eastbourne	9

Fig. 6.2 — WRc intake protection systems: operational and proposed sites in England and Wales — September 1986

(3) A drop in DO from 140% to 40% in under two hours which was partly associated with urban run-off from a nearby town.
(4) An increase in ammonia level from 0.2 to 5.0 mg/l in about one hour.

6.6.2.2 False alarms; alarms which could not be fully interpreted

Some false alarms have been generated by monitoring station maintenance activities, e.g. low frequency vibration. Other, shorter-duration alarms were of uncertain origin but may have been associated with fish disturbance during change-over at the monitoring station concerned, where there is inadequate screening of the duty fish while the other four are being changed.

6.6.2.3 Fish ratio alarms

These have been caused by using weak fish, thereby emphasizing the importance of good fish husbandry, e.g. sound stock of the optimum size (15 cm), adequate feeding (and so avoidance of weak signals), and decreasing the length of the duty period from 14 to 10 days or less in summer temperatures, i.e. above about 20°C.

6.6.2.4 General

No significant mechanical problems have been encountered and some early, minor software faults have been cured. Users have reported that the monitor is sensitive and reliable but no major pollution incident has so far taken place at any of the locations in the short time that the system has been in use.

6.7 BIOMONITORING WITH SENSORS USING MICRO-ORGANISMS AND ENZYMES

The assumption that the response of the fish will be analogous to that of man is better based than it would be for other, simpler organisms and this improves the confidence with which its broadband response is viewed by the user. However, single cells and enzymes hold out the promise of quicker response and, possibly, easier maintenance.

6.7.1 Sensors using micro-organisms

A number of biosensors based on the use of micro-organisms have been used or are proposed for use by the water industry for the detection of pollutants. These sensors are based on the measurement of substrate consumption or product formation.

Holland and Green (1975) developed a sensor system which is based on the inhibition of bacterial nitrification, and designed to respond to abnormal levels of river pollutants. The system consists of a column containing granite chips which is seeded with bacteria which are naturally present in the raw water. The water under test is dosed with ammonia and arranged to flow over the column. Under normal circumstances all the added ammonia will be consumed by the bacteria. However, if a pollutant is present, inhibition of bacterial nitrification will occur. This inhibition is assessed by the measurement of ammonia at the inlet and outlet of the column using an ammonia electrode. Though the detector has been assesed by a water authority (Stroud and Jones, 1972) it never became commercially available. Its major problems include the length of time taken to set up a column (to produce appropriate bacterial flora) and the maintenance required.

Other devices have been reported to be suitable for the detection of pollutants in water and examples are given in the following paragraphs, though these methods are not being used routinely in the UK.

Dorward and Barlsas (1984) developed a sensor for the detection of pollutants based on the use of *E. coli* immobilized onto the surface of a carbon dioxide-sensing electrode. The net rate of carbon dioxide production by the bacteria is a function of the respiratory processes occurring in the cells. The inhibition of any of these processes will result in a decrease in carbon dioxide output by the cells which would be measured by the electrode.

Shieh and Yee (1985) developed a sensor based on the use of bacteria together with measurements of DO. The micro-organisms (obtained from activated sludge) were immobilized onto a biofilter in the form of a biofilm, under oxygen-limited conditions. The system effluent under normal circumstances would contain little or no DO when the micro-organisms are active and the system is free of any inhibitory/hazardous substances. Therefore the magnitude of the effluent DO from the system is a measure of the presence of toxic substances present in the water.

An alternative approach to the measurement of substrate consumption or product formation is to consider the inhibition of electron flow within the electron transport systems of micro-organisms. This holds out the possibility of developing a sensor with a rapid response as well as good sensitivity. WRc has recently let a contract to Luton College of Higher Education, with assistance from Cranfield Institute of Technology, to develop a series of biosensors based on this concept, which would be suitable for use by the water industry for intake-protection purposes.

Chemical mediators can be used to monitor electron transport within selected species of cyanobacteria and eubacteria. This is because in prokaryotic cells the components of the electron transport chains are associated with the outer plasma membranes, allowing easy access to them. Two metabolic events within the cells can be readily monitored: photosynthesis and respiration. Various chemical mediators (e.g. ferricyanide or ferrocene) can be used to capture electrons from the electron transport systems and deliver them to an electrode which is poised at a suitable potential with respect to a silver/silver chloride reference electrode. The magnitude of the resulting current due to the reduction/oxidation cycling of the mediators is proportional to the metabolic activity within the micro-organisms.

Many environmental pollutants of concern to the water industry affect electron transport, either directly by affecting enzymes associated with respiration or photosynthesis, or indirectly by affecting membrane integrity. Depending on the nature and concentration of the pollutant this will result in an acute or chronic disturbance of activity in the electron transport systems. Observation of the magnitude of changes in current resulting from the intermittent stimulation of respiration with glucose or of photosynthesis with light enables the metabolic status of the cells to be continuously monitored.

Two configurations of sensor are presently being developed. In the first type the cells are captured on a filter disc which is then held onto the working electrode, with the mediator solution flowing over the electrode. In the second type of sensor the bacteria are captured in alginate beads held in a column through which the mediator solution passes before it is delivered to a working electrode in a wall jet arrangement.

So far only prototype sensors have been constructed and present research is concerned with the choice of electrode material, type of micro-organisms and mediator to be used. Practical features of the operation of the devices are also being investigated. In particular the life of the sensor may be prolonged by cycling between the working and the resting (or recovery) states of the micro-organisms.

6.7.2 Sensors using enzymes

Enzyme sensors, usually based on the inhibition of an enzyme reaction by a pollutant could be used to detect the low concentrations of pollutants present in water. The ways in which pollutants can inhibit an enzyme reaction are discussed in a review by Tran-Minh (1985).

Enzymes sensors may be expected to be more specific than the whole-cell sensors, though they will still respond to several chemicals within a group. The most successful devices which have been produced to date have been for the detection of pesticides. Goodson *et al.* (1978) produce a sensor for the detection of organophosphate and carbamate presticides based on the inhibition of the enzyme cholinesterase. The level of inhibition, and hence pesticide present, was assessed by adding a substrate and measuring the amount of product by using an electrochemical detector.

Durand and Thomas (1984) produced a similar sensor but assessed the level of inhibition by measuring the pH change following the reaction of the substrate with the enzyme. They were able to detect the carbamate pesticide, carbyaryl, at 1 ppm and the organophosphate insecticide, azinpho sethyl at 1 ppm.

One of the main problems with enzyme sensors relates to immobilization of the enzyme. The immobilization must prevent the enzyme from being washed away, thus extending the life of the sensor, whilst at the same time it must not prevent the pollutants reaching the enzyme. Perhaps more recent developments in immobilization techniques will solve this problem.

6.8 CONCLUSIONS

Because of growing and varied agricultural and industrial pollution there is a need for on-line broadband sensing of pollutants in raw water intended for drinking. Most water undertakings in the UK have no on-line broadband pollution monitoring devices. Any new biosensor to be used in this role must be at least as useful as a fish monitor. It is valuable in considering the relative merits of different types of biosensor to note the following operational requirements:

(1) The device must make measurements at sufficiently frequent intervals — typically at least once every half hour.
(2) It should react to concentrations of pollutants which are dangerous to human level, for which guidelines are available from US and EEC standards.
(3) It should not produce false alarms, defined as alarms for which subsequent GC-MS analysis of the affected water cannot detect pollutants in measurable amounts — at least when GC-MS detection limits are below the recommended

limits. An acceptable rate of false alarms might be once in every two months — a rate higher than this would engender suspicion of the device.

(4) In contrast to medical applications of biosensors, excessive size of a biosensor used in raw water will not be an important issue. Indeed, it may be positively advantageous to use a large sensor if this confers robustness and ease of maintenance.

(5) Biosensors and their associated instrumentation should be easy to install and calibrate by a semi-skilled operator.

(6) Maintenance requirements should be low, preferably less than 30–60 minutes per week (the fish monitor needs 2–3 hours/week). The shelf life of replaceable sensors should be long, so that the water undertaking may carry stock for use as required.

Because biosensing systems which detect inhibition are non-specific, the procedures to be followed in the case of alarm must include further investigation of the nature of the pollutant. For example, an intake protection system controlled by a microcomputer may take samples of the raw water at the time that particular instruments register an alarm. These samples may then be taken with the minimum of delay to an analytical laboratory for detailed analysis, e.g. by gas chromatography–mass spectrometry. Whether or not the intake is closed at the time of the on-line instrument alert will depend on a number of factors. If there is abundant raw water storage, say sufficient for seven days, the decision to close the intake may be taken more lightly, i.e. the consequences of false alarms are less severe. Regardless of the decision-making process which leads to closure, laboratory analysis will be essential to determine when an intake may be re-opened. Throughout this process every attempt should be made to establish the source of the pollution; good liaison will be required with potential polluters and the emergency services.

ACKNOWLEDGEMENTS

The authors wish to thank Dr S. C. Warren, Director, WRc Environment, and Mr R. A. Pepper, General Manager, Sunderland and South Shields Water Company, for permission to publish this paper.

REFERENCES

Cairns, M. A. and Garton, R. R. (1982) Use of fish ventilation frequency to estimate chronically safe toxicant concentrations. *Trans. Am. Fish. Soc.* **111** 70–77.

Dorward, E. J. and Barlsas, B. G. (1984) Acute toxicity screening of water pollutants using a bacterial electrode. *Environ. Sci. Technol.* **18** 967–972.

Durand, P. and Thomas, D. (1984) Use of immobilized enzyme coupled with an electrochemical sensor for the detection of organophosphates and carbamates pesticides, *J. Environ. Pathol. Toxicol.* **5** 51–57.

Evans, G. P. and Johnson, D. (1984) Detection of pollution at drinking water intakes. In T. J. Lack (ed.) *Environmental protection: standards, compliance and costs*, pp. 239–254. Ellis Horwood, Chichester.

Evans, G. P. and Johnson, D. (1985) Automatic detection of pollution at drinking water intakes using fish. *Wat. Supply* **3** 179–186.

Evans, G. P., Johnson, D. and Withell, C. (1986) *Development of the WRc Mk III Fish Monitor: description of the system and its response to some commonly encountered pollutants.* WRc technical report TR233. Water Research Centre, Medmenham, Marlow.

Goodson, L. H., Jacobs, W. B. and Cage, B. R. (1978) A portable pesticide monitor using cholinesterase as the sensor. *Proceedings of a National Conference on Control of Hazardous Material Spills.* April 11–13, 1978. Florida, USA.

Gruber, D. and Cairns, J. (1981a) Data acquisition and evaluation in biological monitoring systems. *Hydrobiologia* **83** 387–393.

Gruber, D. and Cairns, J. (1981b) Industrial effluent monitoring incorporating a recent automated fish biomonitoring system. *Water, Air and Soil Pollution* **15** 471–481.

Holland, G. J. and Green, A. (1975) Development of a gross pollutant detector: laboratory studies. *Wat. Treat. Exam.* **24** 81–99.

Miller, W. F. (1977) *The development of the WRC Biological Monitor*. Paper 8 at a WRc Seminar 'Practical Aspects of Water Quality Monitoring Systems', The Water Research Centre, Medmenham.

Morgan, W. S. G. (1975) Monitoring pesticides by means of changes in electric potential caused by fish opercular rhythms. *Progress in Water Technology* **7** 33–40.

North West Water/Welsh Water (1984) *Joint report on the pollution incident affecting the River Dee and water supplies, January 1984.* Dawson House, Great Sankey.

Rushbrooke, J. N. and Beaumont, F. (1986) River Dee Scheme and intake protection systems. *J. Inst. Wat. Engrs. Scientists.* **40** 173–192.

Shieh, W. K. and Yee, C. J. (1985) Microbial toxicity monitor for *in situ* continuous application. *Biotechnology & Bioengineering* **27** 1500–1506.

Stroud, K. C. G. and Jones, D. B. (1972) Development of a gross pollution detector: field trials. *Wat. Treat. Exam.* **24** 100–119.

Thompson, K. W., Deaton, M. L., Foutz, R. V., Cairns, J. and Hendricks, A. C. (1982) Application of time-series intervention analysis to fish ventilatory response data. *Can. J. Fish. Aquat. Sci.* **39** 518–521.

Tran-minh, C. (1985) Immobilized enzyme probes for determining inhibitors. *Ion-Selective Electrode Rev.* **7** 41–75.

Van der Putte, I., Laurier, M. B. H. M. and Van Eijk, G. J. M. (1982) Respiration and osmoregulation in rainbow trout (*Salmo gairdneri*) exposed to hexavalent chromium at different pH values. *Aquatic Toxicology* **2** 99–112.

Van der Schalie, W. H., Dickson, K. L., Westlake, G. F. and Cairns, J. (1979) Fish bioassay monitoring of waste effluents. *Environmental Management* **3** 217–235.

Van Hoof, F. (1980) Evaluation of an automatic system for detection of toxic substances in surface water using trout. *Bull. Environm. Contam. Toxicol.* **25** 221–225.

Wallwork, J. F., Heslop, R. T. and Redshaw, N. L. (1977) A river monitoring station for protection and control of water treatment. *Prog. Wat. Tech.* **9** (5/6) 215–220.

Wallwork, J. F. (1980) Water supply intake protection — river water data collection and pollution monitoring. In M. J. Stiff (ed.) pp. 175–187. Ellis Horwood, Chichester.

Wallwork, J. F. and Ellison, G. (1983) Automated sensor systems for water resource pollution warning and treatment process control. *Aqua* **6** 313–320.

Water Authorities Association/Water Companies Association, UK. (1984) *Actions to minimise the effect of pollution incidents affecting river intakes for public water supplies.* 25pp. London.

7

Effluent discharge control at a South African industrial site utilizing continuous automatic biological surveillance techniques

W. S. G. Morgan
National Institute for Water Research

P. C. Kühn
National Electrical Engineering Research Institute

Council for Scientific and Industrial Research, PO Box 395, Pretoria 0001, South Africa.

7.1 INTRODUCTION

The quality of human life depends upon progressive industrial technological advancement; human survival upon the maintenance of a relatively stable natural environment. The two are not necessarily compatible. Industry creates waste in relation to the prevailing market demand for its products rather than to the capacity of the environment to accommodate that waste. However, increasing public concern over the impact of pollution on the environment has resulted in industry being compelled to evaluate meticulously the influence its wastewaters have on biological systems. In this respect, industry is, understandably, concerned that viable techniques be employed to provide valid assessments at minimum cost. Research personnel, therefore, bear a responsibility to provide techniques which will satisfy both industrial and environmental requirements.

At present, control and surveillance of industrial wastewater discharges depend largely upon physicochemical measurements which, because of the discontinuity in sampling procedure, do not always detect damaging effluent discharges in time. Furthermore, such measurements cannot predict the toxicity of complex wastes where the components may act synergistically or antagonistically with one another. A biological monitoring system, functioning in continuous mode, is not only able to integrate these effects, but also to provide an advance warning of insiduous toxic conditions.

Biological systems employing fish physiological and behavioral responses to monitor toxic effects have been under development at the National Institute for Water Research for a number of years (Morgan and Kühn, 1974; Morgan, 1977a,b). Laboratory tests have proved the feasibility of using locomotor behavior patterns of fish to assess pollution (Morgan, 1979; Shirer *et al.*, 1968).

An improved automated biomonitoring system, based upon fish activity responses to intoxication, has recently been installed at a pulp and paper mill at Mandini in Natal, South Africa, with the co-operation and financial support of Sappi Kraft Ltd. This chapter describes the application and operation of the system and discusses future improvement for in-plant exploitation, based upon practical experience.

7.2 THE BIOLOGICAL MONITOR

Waller and Cairns (1972) found that fish movement patterns, monitored by light-beam interruption, were a reliable index of pre-mortal toxicity. However, equipment employing light-beam interruption to monitor locomotory behavior is limited by turbidity and, therefore, inappropriate for use in an industrial situation. A similar system used in our laboratories (Morgan, 1979) required regular realignment of light emitters and receivers after slight vibrations. An ultrasonic technique for monitoring activity was found to be free of these drawbacks.

The monitoring equipment was developed as an on-line, automatic, continuous biological system to detect potentially harmful substances in flowing water (Fig. 7.1). The system can monitor either five or 10 independent channels, depending upon whether one or two five-channel monitoring chassis are fitted. Each channel consists of a specifically designed sensor chamber containing one fish, through which passes a continuous flow of test water. The system operates by monitoring and recording the activity level of each free-swimming sensor fish. The activity of each channel is also checked against the adjustable critical threshold. Should the activity of the sensor fish exceed this level, a response for that particular channel is recorded. Whenever any three of five sensors elicit a response, an external audible alarm signal is generated. The data for each channel are recorded automatically on a printout once per hour, or manually on demand.

The system monitors the activity of the sensor fish by means of ultrasonic echoes. Each sensor chamber is equipped with one ultrasonic transmitting element and one receiver (Fig. 7.2). The signal received comprises the sum of the various reflected sound waves. Whenever the fish moves, those reflections coming from its body exhibit a Doppler shift in frequency which, after mixing in the receiver with the other reflections, appear as an amplitude modulation of the total signal. The modulation is processed and used to operate the activity counters, which advance whenever the fish moves.

The operational features of the system are depicted in Fig. 7.3. Ten tanks are each fitted with a pair of ultrasonic transducers, one acting as transmitter, and the other as receiver. The transmitters are driven by a 100 kHz sine-wave signal. The receiver in each tank reacts to the sum of the reflected signals, a proportion of which emanate from the body of the fish. Movements of the fish result in Doppler frequency shifts in the received signals. The signals from each tank are passed to separate signal processing modules, where they are amplified utilizing automatic gain control,

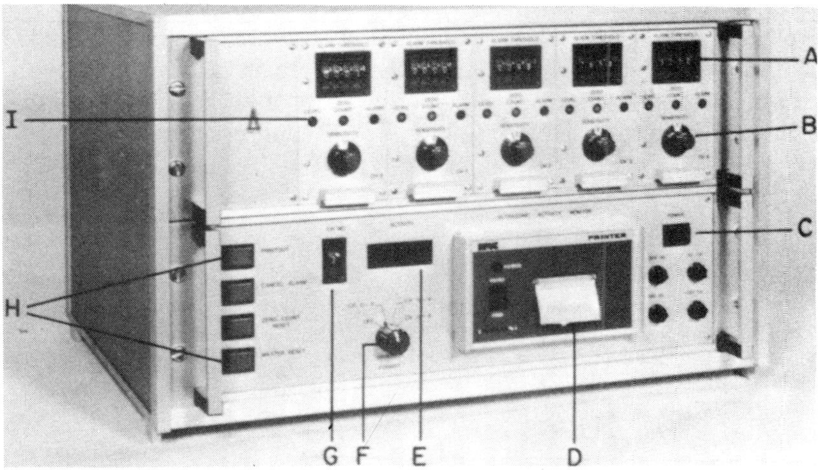

Fig. 7.1 — Five-channel electronic monitoring unit. A, four-digit activity threshold control; B, sensitivity control; C, power ON/OFF switch; D, printer; E, four-digit light-emitting diode (LED) activity count display; F, channel printout selector switch; G, channel select control; H, manual printout, alarm and master reset controls; I, signal level and alarm indicator lights.

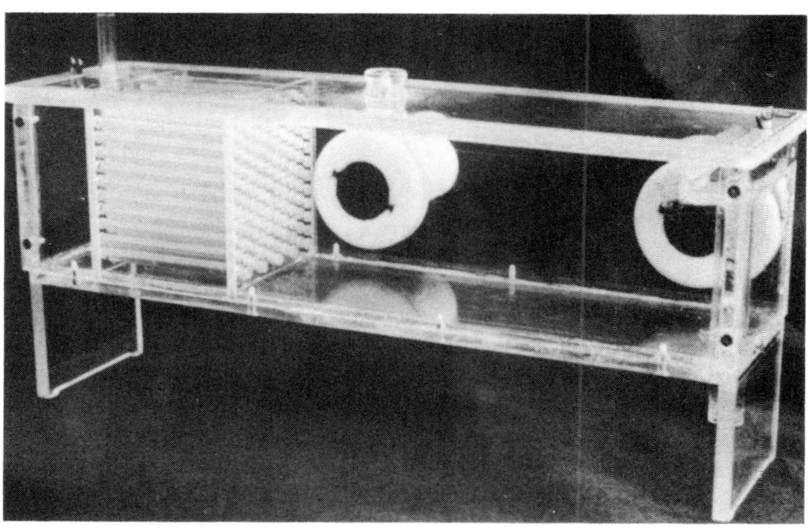

Fig. 7.2 — Activity sensor chamber showing ultrasonic emitter and receiver.

before being filtered and detected. A pulse is generated whenever the movement signal of a channel exceeds a discrimination threshold, set via an independent sensitivity control on each channel module. The rate of these pulses thus provides a measure of the fish's activity level. The duration of each pulse is set at 1 s, thus providing a 'pseudo full-scale' sensitivity of 3600 counts per hour.

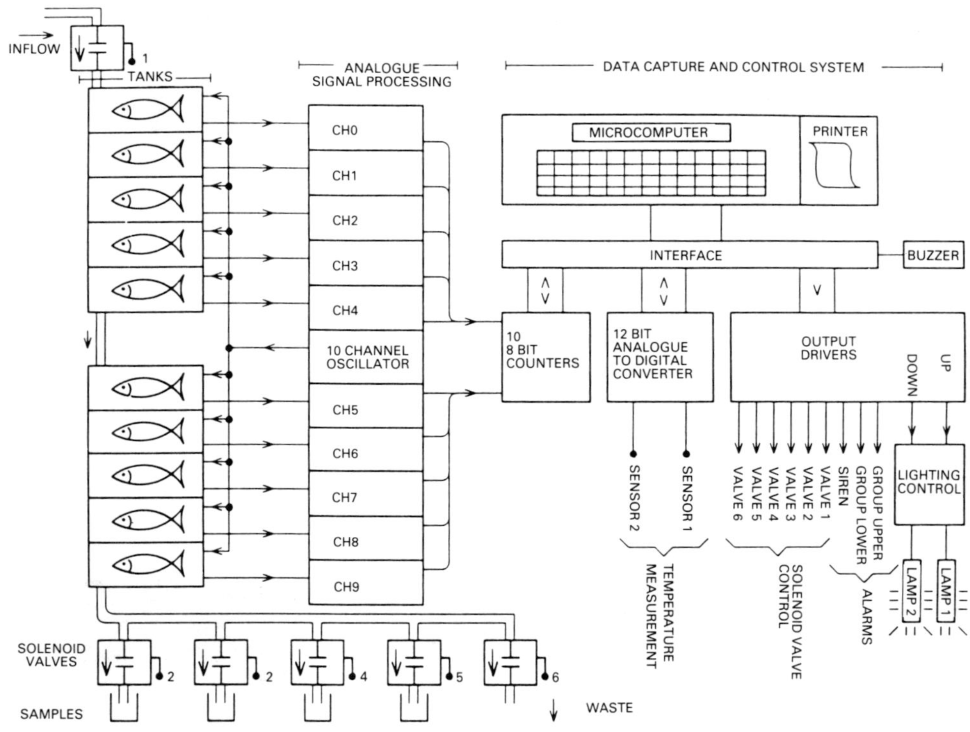

Fig. 7.3 — Functional diagram.

A significant design feature is incorporated in the transmitting transducers. These devices have a nominal resonant frequency of 100 kHz, but in practice exhibit deviations in this parameter of up to 5%, owing to manufacturing tolerances. In previous systems of this type, sets of transducers were selected for minimum spread in characteristics, and the oscillators had to be carefully tuned to accommodate these fluctuations. The circuitry in this system has been designed so as to utilize a single frequency-modulated oscillator, of which the output frequency is continually 'swept' between 90 kHz and 110 kHz at a rate of 120 Hz. This approach ensures that the drive signal will pass through the resonant peak of each transducer, no matter what its precise value, thus providing a stable output. This strategy has proved most successful in so far as the system requires no selection of transducers or adjustment, and is impervious to transducer tolerances, aging effects and ambient temperature changes.

The data capture and control functions fall into six categories as follows:

- Counting of the activity pulses from the 10 channel modules.
- Analogue-to-digital conversion of the signals from two temperature sensors, used for correlating the effect of water temperature on fish activity.
- Keyboard entry of certain command parameters, and display of data and operator prompts.
- Manipulation and analysis of the measured data and print-out of results.
- Control of two 'Group alarm' indicators and one 'General alarm' siren.
- Control of six solenoid valves, which divert the water flow under toxicant conditions, so as to prevent desensitizing of the fish sensors, and at the same time sample the water automatically, for subsequent analysis.

These functions were all realized through the use of a programmable data acquisition and control unit developed at our laboratories for general use in a research environment. The unit is based on a microcomputer with keyboard and liquid crystal display (programmable in BASIC), a four-colour printer/plotter and a cassette tape-recorder for program and data storage. The unit features non-volatile memory, and a real-time clock/calendar is also included. The unit was customized by the inclusion of a 12-bit analogue-to-digital convertor (for temperature measurements), three four-channel counters (for activity counting) and nine output switching modules of solid-state relays (for solenoid valves, lighting control and alarm generation).

When the system is switched on and the individual sensitivity controls set, further operation is executed automatically. The activity pulses from the analogue modules are accumulated in the 10 counters. These are read every six minute (in two three-minute segments to avoid possible over-range errors in the counters) and stored in an array in memory, thus providing 10 figures for each channel per hour (Fig. 7.4).

At the end of each hour, this data is used to calculate an hourly mean and standard deviation figure for each channel, which are stored in a separate array. Confidence indexes are calculated from these and compared with those of the corresponding hour in the preceding 24-hour period. If the response criteria are not exceeded, the existing confidence indexes are replaced, and the procedure repeated for the next hour. If a response is detected, however, the existing index data are retained. This strategy provides an adaptive decision model which requires no lengthy initial calibration procedure, in contrast to preceding systems of this type. Two temperature sensors are also read and an hourly mean temperature calculated in each case. An hourly print-out is generated at this point, giving the relevant data for each channel and their response status, together with the mean temperatures.

If three or more out of either group of five channels have exhibited responses, the associated 'Group alarm' indicator is illuminated and the 'General alarm' output is triggered, causing a buzzer to sound. If this is not canceled within one minute, a powerful siren is activated. At the same time, the solenoid valves are activated according to the following strategy: the wastewater component of the mixture flowing through the fish tanks is diverted and the tanks are flushed out with the clean water component. The wastewater component is sampled automatically by means of

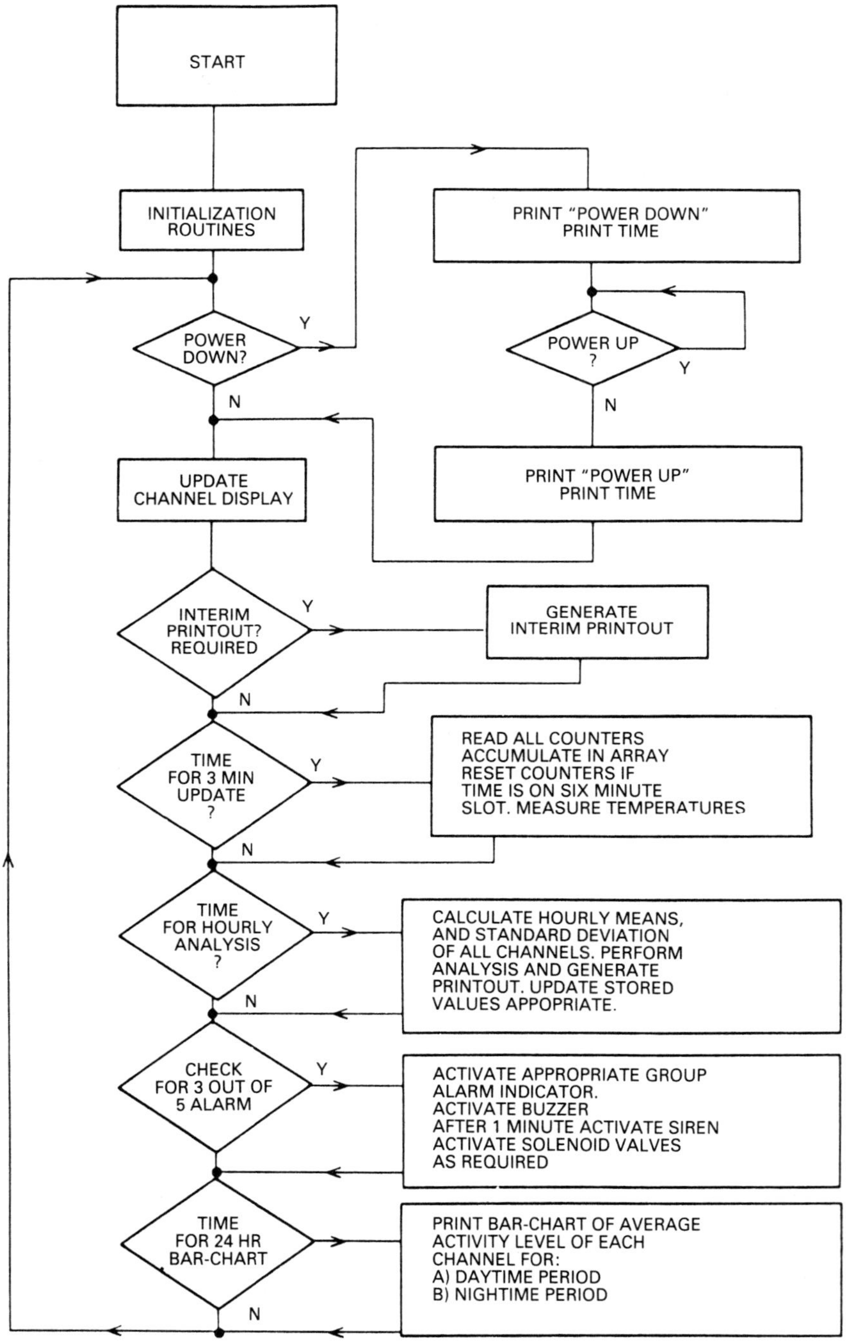

Fig. 7.4 — Simplified software flowchart.

another solenoid valve for a fixed time, after which it is switched to waste. At the end of the next hour the normal flow is restored. If, after another hourly period, the alarm situation is again detected, the whole alarm sequence is repeated; otherwise normal operation continues.

Since the tanks are all totally enclosed, it is necessary to create an artificial day/night cycle. This is done by means of incandescent strip lights mounted inside the tank enclosures. These are also under computer control, via a hardware current-control circuit. The light intensity is ramped up over a half-hour period at a preset time (usually 0615 h), and are similarly dimmed again in the evening (normally 1815 h).

Once in every 24-hour period (currently set at 0900 h), a bar chart representing the mean activity levels for each channel during the day and night periods is printed out.

Protection against unexpected power outlays is provided by a 'POWER DOWN' detection circuit, as well as by the non-volatile nature of the program and data memory, and by the real-time clock. Monitoring of activity levels naturally ceases for the duration of the power failure, owing to the interruption of the analog circuitry, but the integrity of both the existing data and the program is preserved. When a power failure is detected, a 'POWER DOWN' message is printed, together with the time at which the power failure occurred. When power is restored, a similar 'POWER UP' message is printed, and system operation is automatically restored.

The system can also be interrogated from the keyboard at any time so as to display the current activity level of any channel, or to provide an interim print-out of all channels.

7.3 IN-PLANT APPLICATION

The Tugela River mouth lies midway between Durban and Richards Bay (Fig. 7.5). It is 10 km from the National road and 104 km from Durban. The biggest integrated pulp and paper mill on the African continent, SAPPI KRAFT LTD, is sited approximately 200 m from a bend in the Tugela River immediately above the estuary. The mill uses 970 000 tons of timber annually, and produces 288 000 tons of finished paper per annum. The Kraft cooking process for pine wood is used, therefore the isolation of the site has the advantage that odours produced by this process do not create a nuisance to the large urban population further south. Apart from the Kraft process, gum wood is pulped by a neutral sulphite semi-chemical (NSSC) process. The latter produces a lignin-rich liquor which is disposed of safely by combustion in a fluidized-bed Copeland reactor.

Approximately 48 000 m^3 of water are pumped daily from the Tugela River, of which 10 000 m^3 are used for domestic purposes, whereas the remainder is used for pulp and paper making (Fig. 7.6). The mill effluent treatment system consists of a round, sectioned clariflocculator for settling the fibrous wastes from the paper machines, and a rectangular clarifier for treating the overflow from the clariflocculator and a separate chemical process stream. The purpose of both clarifiers is to promote separation and no secondary treatment is undertaken. The biomonitoring facility was sited at the outlet of the rectangular clarifier in order to monitor the final mill effluent before it flows into the Mandini stream. The Mandini also receives

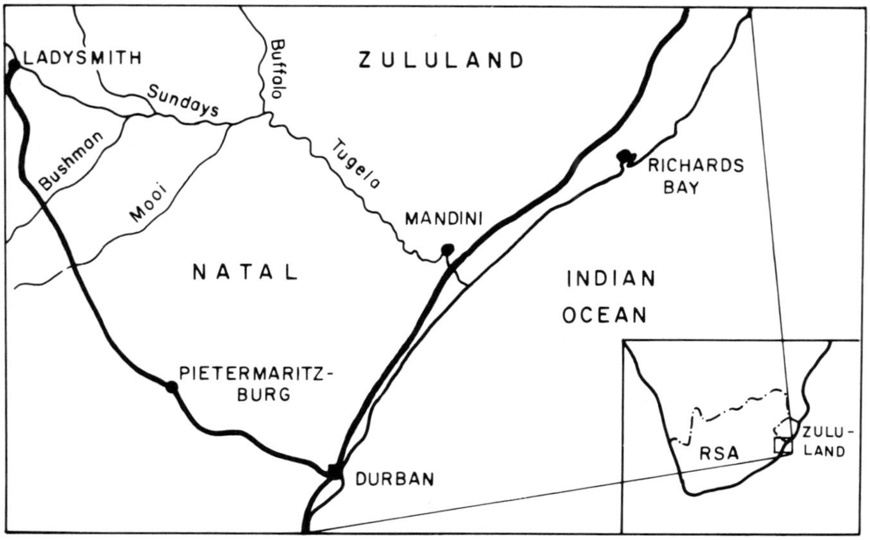

Fig. 7.5 — Location of SAPPI KRAFT LTD, Mandini.

effluent from the Isitebe industrial estate and sewage effluent from the Sundumbeli treatment works. Both effluent discharges enter the Mandini upstream of the mill. The objectives of biomonitoring were to:

- give warning of discharges of toxic chemicals;
- assess the need for secondary treatment;
- provide early warning of possible permit violations;
- protect the aquatic environment;
- promote better operational functions within the mill; and
- monitor the Mandini upstream of the mill.

Three five-channel biomonitoring modules were established in a laboratory designed in such a way that chambers containing the sensor fish were physically separated from the electronic units. Thus the fish experienced the minimum of disturbance from operating personnel. Sensor fish in chambers linked to one of the three modules acted as controls, receiving a continuous flow of dechlorinated mill water. Those maintained in a second set of chambers received mill effluent diluted to a concentration equivalent to that pertaining at the confluence of the Mandini stream and the Tugela River at the lowest recorded river flow. Examination of river flow-rate records spanning a period of 10 years revealed a minimum river flow-rate of 590 000 m^3/day (6.83 cumecs), thus affording Tugela Mill's discharge of approximately 30 000 m^3/d with a dilution factor of 20 : 1. The diluent water flow was regulated to give that factor and controlled by analysing the combined effluent and mill water for its sodium content. A separate five-channel unit was employed to monitor water extracted from the Mandini stream above the mill discharge point.

The fish species used as sensor was *Poecilia reticulata* (guppy). This species was

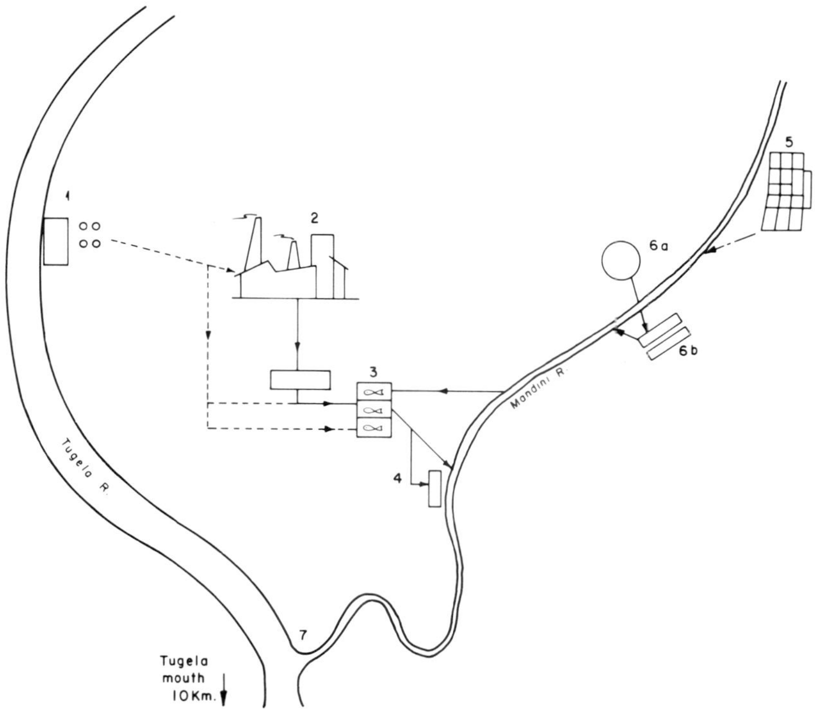

Fig. 7.6 — Water flow through the mill and fish biomonitoring system at SAPPI KRAFT LTD (Tugela Mill) in relation to other possible sources of pollution. 1, Pump house and clarification ponds; 2, SAPPI KRAFT (Tugela Mill) LTD; 3, Biomonitoring laboratory; 4, Emergency dam; 5, Isitebe industrial estate; 6, Sundumbeli sewage works; 7, Confluence of Tugela River and Mandini stream.

selected because it is more sensitive to a wide variety of pollutants than any of the species indigenous to the Tugela River, and can easily be maintained and reared in a laboratory environment. A constant supply of fish are readily available from original stocks, as females produce live young about once a month. This breeding propensity, however, precluded the use of female fish as sensors. More than one fish per chamber would provide anomalous activity levels.

7.4 RESULTS

There has been a decided improvement in effluent quality during the 12-month period that the system has been in operation (Table 7.1). This has been mainly due to the incorporation of an additional rectangular clarifier as part of the effluent treatment process, which was indicated as being necessary during preliminary biomonitoring tests. However, the realization that a continuous effluent monitoring system is in operation at the mill has also induced personnel to become more diligent with respect to waste-disposal practices.

Only one fish kill has occurred in the Tugela River during the 12 months that the

Table 7.1 — Analysis of mill effluent six months before and six months after the installation of the biomonitoring system*

Determinand	Before	After
Chemical oxygen demand (mg/l)	431 (250–780)	260 (150–470)
Sodium (mg/l Na^+)	215 (179–237)	84 (70–95)
Suspended solids (mg/l)	87 (32–150)	42 (14–106)
Dissolved solids (mg/l)	945 (876–1113)	842 (607–1023)
Colour (Hazen units)	172 (110–290)	140 (90–230)
pH	7.8 (7.6–8.0)	7.8 (7.5–8.0)
Dissolved oxygen (mg/l)	2.5 (1.91–3.12)	3.8 (3.1–4.3)
Calcium, Ca^{2+} (mg/l)	39.9 (26.5–54.3)	37.8 (31.7–51.0)
Magnesium, Mg^{2+} (mg/l)	15.7 (9.9–17.3)	12.5 (6.4–16.2)
Conductivity (μS/m)	1038 (960–1250)	920 (760–1090)

*Figures are averages for the whole six-month period, with the range in brackets.

biomonitoring system has been in operation. Investigation by regulatory and environmental agencies of the activity records obtained from fish maintained in the sensor modules indicated that the source of pollution, in this instance, was upstream of the mill's discharge point. All five sensors exposed to Mandini stream water elicited responses on the day preceding the fish kill. The elapsed time between responses obtained from the most sensitive and the most resistant sensors was approximately 12 hours, indicating the effects to be acute. On receipt of the responses from 50% of the sensor fish, indicative of an alarm condition, control water was diverted to the sensor chambers. Although this did not prevent the other sensor fish from responding, only one of the five sensors died. Future occurrences of this type may, therefore, be alleviated by the mill, although not responsible for the pollution incident, diverting process water to the Mandini stream to moderate the toxic effect?

In this instance the industry were able to prove that they were not responsible for the fish kill, thereby avoiding possible litigation. It is understood, however, that management must be prepared to acknowledge deleterious situations arising from their own industrial spills — otherwise the whole biomonitoring exercise would lose credibility.

7.5 COSTS OF BIOMONITORING

The total cost of installing the biomonitoring system at the mill was approximately R35 000 (R1=US$0.44). Most of this was accounted for by labour and building costs for the construction of the laboratory, and provision of ancillary services. At the time of installation, the cost of hardware for the electronic monitoring and alarm units was approximately R7500, with each sensor chamber costing R200 to have made. It should be noted, however, that this relatively low expenditure resulted in a saving to the company of between R3 and R5 million, since it could be indicated that second-stage effluent treatment, as originally required by the regulatory authorities, has, to date, been unnecessary in this geographical situation.

The essential automatic and continuous nature of the biomonitoring system entails minimal operating and maintanance costs. Running expenses for the first twelve months of operation totalled R3350, a sum which compares favourably with the cost of R3500 incurred by the company in carrying out physicochemical analyses of the effluent.

7.6 ENVIRONMENTAL AND INDUSTRIAL BENEFITS

Cairns *et al.* (1973) stated that ‘fish sensors in continuous monitoring units at industrial sites can warn of developing toxic conditions in time to forestall damage to the fish populations in streams’. This aspect of effluent quality control has been enhanced, at the Tugela mill, in two ways. Firstly, the sensor fish utilized for monitoring are more sensitive to toxic substances than those indigenous to the river, and, secondly, the most critical effluent concentration level is assessed, the average dilution factor for mill effluent being 100 : 1 rather than the 20 : 1 dilution tested in the monitoring facility.

The benefits accruing to the industry from the installation of the system are more difficult to assess. Managements are, understandably, reluctant to become involved in what is, after all, an explorative development, with the attendant possibility of additional restrictions being imposed by governmental and regulatory authorities. The fact that, in this instance, preliminary experimentation using the system indicated a level of effluent treatment financially acceptable to the company was decisive in this respect. This will not always be the case. In many instances, more expensive treatment may be indicated. However, industry, investing in continuous monitoring systems for water quality control may be recompensed by the positive publicity and enhanced public relations that the use of such systems would bring about.

Industry also stands to gain from rapidly detecting losses of valuable substances. At the Tugela mill, for example, recovery of inorganic chemicals from the Kraft process is an economic necessity, whilst the NSSC process yield marketable by-products. Laboratory simulations of process disfunction have indicated that any alteration in the characteristic nature of the effluent, owing to malpractice or malfunction, would be detected biologically in a very short time.

7.7 FUTURE DEVELOPMENTS

The biomonitoring system described has functioned adequately throughout its first 12-month operation period, and served the main purpose for which it was designed (the detection and control of acutely toxic materials which would have deleterious effects upon the receiving ecosystems). This does not mean, however, that all functional parameters have been assessed, or that no further improvements are necessary.

Whether the in-plant biomonitoring system detects effluent concentrations which may have a long-term chronic effect upon fish in the receiving water has not been established. Laboratory experimental investigations into the effect of mill effluent and its components upon growth and reproduction in sensor fish have been initiated. Should these parameters be inhibited by a lower concentration than that to which sensor fish effectively respond within the biomonitoring system, then a relative safety factor would need to be introduced by metering a proportionately higher effluent concentration to the sensor fish than is delivered to the river.

The natural biological delay in response to toxic situations is, at present, aggravated by the fact that effluent is manually diverted, on receipt of an alarm, to the emergency facility. This procedural delay would be minimized should the electronic monitor itself, by means of signals transmitted to automatic valves in the waste line, undertake this function. This modification is currently being evaluated.

Finally, it is expected that an on-going assessment of possible synergistic and/or antagonistic interactions between the mill effluent and upstream industrial and domestic discharges will be of value in setting criteria for future development in the area.

7.8 CONCLUSION

The adoption of a biological monitoring system at SAPPI's Tugela mill, as a major factor in the determination of effluent discharge specifications, represents a radical shift in the approach to environmental quality control by regulatory agencies. Whether biomonitoring systems gain wider acceptance depends to a large extent, upon the performance of the unit at Mandini. A great responsibility, therefore, lies with the personnel involved to justify its more extensive application upon sound scientific and operational principles.

The catchment area of the Tugela River system is one of great natural beauty, but also one ripe for industrial and commercial development. We believe that the incorporation of automatic biological monitoring systems into the water quality control complex will preserve the former without prohibiting the latter.

REFERENCES

Cairns Jr, J., Sparks, R. E. and Waller, W. T. (1973) The use of fish as sensors in industrial waste lines to prevent fish kills. *Hydrobiologia* **41** 151–167.

Morgan, W. S. G. (1977a) An electronic monitoring system to monitor the effects of changes in water quality on fish opercular rhythms. In J. Cairns Jr, K. L. Dickson and G. F. Westlake (eds) *Biological monitoring and effluent quality,* ASTM STP 607. pp. 38–55. American Society for Testing and Materials.

Morgan, W. S. G. (1977b) Biomonitoring with fish: an aid to industrial effluent and surface water quality control. *Prog. Water Technol.* **9** 703–711.

Morgan, W. S. G. (1979) Fish locomotor behaviour patterns as a monitoring tool. *J. Wat. Pollut. Control Fed.* **51** 580–589.

Morgan, W. S. G. and Kühn, P. C. (1974) A method to monitor the effects of toxicants upon breathing rate of largemouth bass (*Micropterus salmoides* Lácèpede). *Wat. Res.* **8** 67–77.

Shirer, H. W., Cairns Jr, J. and Waller, W. T. (1968) A simple apparatus for measuring activity patterns of fishes. *Wat. Resource. Bull.* **4** 27–43.

Waller, W. T. and Cairns Jr, J., (1972). The use of fish movement patterns to monitor zinc in water. *Wat. Res.* **6** 257–269.

8

Avoidance reaction of fish to an industrial effluent and its constituent toxic components

J. Hadjinicolaou and **L. D. Spraggs**
Department of Civil Engineering and Applied Mechanics, McGill University, 817 Sherbrooke Street West, Montreal, Canada H3A 2K6

8.1 INTRODUCTION

This paper examines the acute avoidance behavior of rainbow trout to sublethal concentrations of industrial effluent.

A toxicant may stress fish populations in ways other than acute mortality, but only recently has substantial attention been given to sublethal effects. Field studies have shown that toxic solutions may render large portions of an environment undesirable to fish (Birtwell, 1977). The behavior of fish in these areas is often modified; any behavior patterns that allow fish to escape or to avoid deleterious conditions would contribute to their survival and well being (Cherry and Cairns, 1982).

Among behavioral tests, test of preference–avoidance reactions are particularly important and practical because they can detect adverse effects of pollutants at sublethal concentrations. These tests are of relatively short duration (normally a matter of hours), and their outcome is measured by the alteration of behavior, rather than by effects on particular organs.

Avoidance curves and thresholds might provide valuable insight regarding the behavior of fish toxic solutions that cannot be obtained from static or chronic bioassays.

Three basic types of apparatus have been used to analyze chemical avoidance. In 1913, Shelford and Allee (Shelford *et al.*, 1913) first tried shallow concentration gradients. Wells (1915), Ishio (1969) and Birtwell (1977) used and modified this technique. Jones (1947) introduced the steep gradient method. Finally, the fluviarium system was introduced by Höglund in 1951. Other pioneers notably Sprague (1964), Kleerekoper (1967), Scherer and Nowak (1973), and Westlake and Lubinski (1976) have all used modifications of the above three methods.

The system outlined in this paper was designed to take advantage of the success of previous researchers (to use steep gradient design rather than shallow or fluviarium system, the ability to store results for later re-examination, to use actual effluent instead of only purified chemical components) and at the same time to provide possible new advantages (to have vertical, longitudinal and lateral analysis of the organism's position, to increase the number of test organisms, the use of a large tank which eliminates end effects).

8.2 MATERIALS AND METHODS

The entire design of the new flow-through system (Spraggs *et al.*, 1981, 1982; Hadjinicolaou and Spraggs, 1981, 1982) can be divided into five basic components:

(1) the water treatment facilities,
(2) the holding facilities,
(3) the temperature control system,
(4) the channel,
(5) the data acquisition facilities.

Fig. 1 depicts the system schematically. A number key below the diagram indicates its constituent parts.

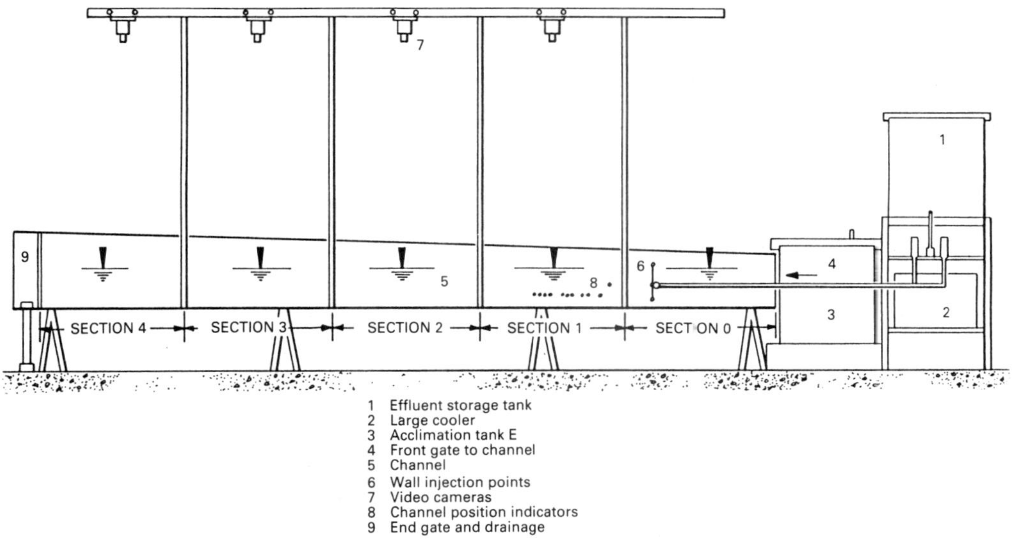

Fig 8.1 — Side view of avoidance apparatus.

Water for the system was taken from the domestic supply of the City of Montreal. Prior to actual use, it was treated to render it suitable for fish survival. The water treatment facilities were an activated carbon unit and ultraviolet lights to kill bacteria. The maximum amount of treated water was 20 gpm.* Water analyses were done monthly to evaluate the water quality before its entrance into the holding facilities. (Averages of measured values: pH=6.3±0.3, temperature 15°C±1°C, sodium 14.2±0.8 ppm, calcium 23.5±2 ppm, chromium <0.002 ppm, nickel <0.002 ppm, iron <0.002 ppm, total and fecal coliforms 0.)

The holding facilities were the storage reservoir, for the cleaned water (tank A, 500 gallons), the fish tanks (B, C and E, 300 gallons) and tank D as a storage reservoir for industrial effluents. Flowmeters and airmeters regulated the amount of water and air that flowed into the tanks and the channel. Tank D was mounted on a 2 m-high steel support structure and the industrial effluent was pumped through one of two flowmeters to give a constant flow of effluent into the channel. On the one side of tank E a 0.4×0.3 m hole was made, 0.25 m from the top of the tank, to allow the fish to move from the tank E to the channel through the front gate.

A submerged heating unit in tank A and an automatic temperature control unit provided a continuous temperature of 15°C in the system. In addition, cooling units were used to cool the water to 15°C in the summer. Ten waterproof temperature probes were strategically located throughout the system (channel, holding tanks, etc.) and connected to a manually operated multi-channel thermometer for rapid recording of the temperature at any point in the system. The maximum temperature differences between different sections of the channel stayed within 1°C.

The experimental channel was 10 m long, 0.305 m wide and had a water elevation of 0.35 m. It was divided into five sections of 2 m each. The first section was constructed of plexiglass to allow for holes for jets and diffusers, whereas the other four were made of glass. In the first section of the channel, 1.7 m from the front gate and 0.3 m from the second section, provision was made for the injection of the effluents and dyes from the effluent tank D. The flow of the channel was laminar, with a rate of 6 gpm. The injection system used in this study was a wall momentum-jet 10 cm from the bottom.

A 8 mm-thick plexiglass barrier with holding devices was placed in the centre of the channel extending from 1.35 m upstream of the first section through to the third section (4.65 m long). This separation unit rose 0.1 m above the water level and was introduced to give the fish a chance to choose either the polluted side (polluted sections 1,2,3 and 4) or the non-polluted side (section 0, or non-polluted sections 1,2,3 and 4) of the channel. The fish could not go over or under the separation unit but only around it, from sections 0 and 3.

Because it was desirable to have a longitudinal, vertical and lateral analysis of the disposition of the fish in the channel, mirrors were placed along one side of the channel. The mirrors were placed at an angle of 45° and, when photographed from above, showed the longitudinal projection in the direct view and the vertical projection in the mirror. Consequently, any avoidance in either the lateral, longitudinal or vertical direction was recorded. Direct longitudinal coverage was provided by four overlapping cameras, mounted on moveable trolleys 3 m above the channel and along its length (Fig. 8.1). Each camera can cover approximately 2.2 m of the channel, resulting in total coverage of the area where the fish are able to swim. The cameras project on to the monitors: (a) an overhead view which provides observation of the longitudinal and lateral distribution of fish within the section and (b) a

*US gallons are used in this chapter. 1 gallon=3.785 litres.

side view through the 45° mirrors which provides information about the vertical distribution of fish within the section.

Simultaneously with the pictures, long-playing video recorders are used to record the video signals for later analysis. The four video cassette recorders store the events of every experiment, both vertically and longitudinally. Pictures from the four cameras are displayed on a separate video monitor for each section of the channel.

The image in each monitor is comprised of two parts: the upper part which corresponds to the vertical distribution (depths A, 0–0.11m; B, 0.11–0.22 m; C, 0.22–0.33 m) of the particular section, and the bottom part which illustrates the longitudinal distribution (sections 1, 2–4 m, 2, 4–6 m; 3, 6–8 m; 4, 8–10 m) and the lateral distribution (sides A and B) within the particular section.

The bottom and one side of the entire channel were divided into 0.101×0.101 m squares with black stripes. Each of these squares was color-coded horizontally and vertically to aid in determining the position of the fish in the channel. Each section had 18 squares.

Because the light reaching the iris of the cameras was not adequate and because the picture on the monitors from the mirrors was not clear, fluorescent lights 1.2 m long were placed along the length of the channel 5 cm from one side. These lights had a 12-hour photo period, with lamp strength of 2×34 W and were covered by a plastic shield.

Large portable modular wall panels were located along one side of the channel to minimize both noise and visual disturbances while experiments were in progress.

A plexiglass box with a perforated end gate was built at the downstream end of the channel. The perforated end gate was required to allow for flow which approximated parallel flow in a cross-section of a river while at the same time holding the fish within the channel. The diameter of the holes (4 mm) and length of slots (7.5 cm) in the end gate prevents small fish from passing through. The end gate was removable for rapid drainage and easy cleaning of the channel, which was performed daily and had to be completed eight hours before beginning an experiment, by flushing three times with water containing a very low (1 : 60) versa-clean liquid concentrate dilution.

The species used in the study was the rainbow trout (*Salmo gairdneri, R.*) from two Quebec hatcheries, originally 2–3.5 g in weight (5–7 cm in length), with a growth rate of approximately 2 cm per month. The fish food used was trout food-Atlantide, size 3. There were no mortalities during holding or experimentation except once, caused by a power failure. The fish were kept in tanks B and C until they became 10–11 cm in length before being transferred to the channel acclimation unit (tank E), to acclimate the fish to living either in the channel or in tank E.

Fish feeding was done at the same time each day at five specific locations in the channel to avoid grouping. No experiments began until three hours after feeding. Fish had to be counted prior to the commencement of any experiment. The count took 30 minutes and was done by comparing the number of fish on five still frames of a continuous 10-minute recording, together with a corresponding visual count of the number of fish in section 0.

The temperature of the effluent and the channel were checked every half-hour. The effluent was discharged for 15 minutes before actual recordings began. A switch box was connected to the four cameras. By turning the switch on, the experiment was simultaneously recorded on the four cassette recorders for two hours continuously. Any changes in the ambient conditions had to be recorded and any resultant behavior changes likewise noted. The experiment was supervised during its entirety including changes in personnel. After two hours, the experiment was terminated by turning the switch off.

Following any experiment, there was a series of procedures which were followed to ensure that subsequent experiments were not biased. Cassettes were rewound and temporarily labelled. Fish which had been exposed were removed and destroyed:

— no fish which had been exposed to the effluents were placed back in the acclimation tank E or re-used;
— no fish which had been moved to the experimental tank were returned to the stock tanks.

The channel was drained, refilled and fish were allowed to enter the channel for acclimation for subsequent experiments.

Because 100 fish were used in each experiment their position in the channel had to be noted at selected time intervals.

This analysis required an initial visual count followed by an analysis of the position of the fish. To determine the relative avoidance of the fish, the longitudinal, vertical and lateral position of the fish was determined at each counting interval (5 min).

From the counting sheets the data were transferred to computer files and the statistical analysis consisted of preliminary calculations (mean value, standardization, range and percentage of the number of fish per section), time series (to evaluate the stability of the response with time on the sides A and B of the lateral distribution), analysis of variance (to evaluate the likeness of having significantly different values among the percentages of fish present in each section and at each concentration), and Duncan's tests (to analyze which value on a group of data differed from the rest, after it was found from ANOVA that the values were likely to be significantly different). To ensure that the results were not biased, the tests were conducted with effluent being inserted in alternate sides of the channel for alternate tests.

To obtain the longitudinal distribution curve of a particular experiment, a transformation was required of the total number of fish per section into percentages, and the plotting of these values versus the four longitudinal sections.

To obtain the vertical distribution curve for a particular experiment, a transformation was required of the total number of fish per depth into percentages and the plotting of these values versus the three depths.

In the lateral distribution the main avoidance was defined as the mean percentage of fish in the non-polluted side of the first two sections. Secondary avoidance was defined as the mean percentage of fish in the non-polluted side of the third and fourth sections, due to the difficulty of deciding about the orientation of fish which were observed between the polluted and the non-polluted sides of these sections. In these calculations the first four readings were excluded because the first 20 minutes were evaluated as an adaptation period in the channel. The main avoidance after the elimination of the bias factors (water jet effect) represented the avoidance value of the experiment and was used for the production of the avoidance curve. The secondary avoidance was used for additional information.

To obtain the avoidance curve, the time spent in clear water as a percentage of total time was plotted versus the effluent dilution concentration.

Generally in the avoidance curves each experiment is represented by a single value, which is the percentage of total experimental time spent by the fish in clear water. Avoidance reactions have usually been portrayed in this fashion, and only rarely have attempts been made to use other parameters. However, in order to provide more insight into the meaning of this single value, it was felt worthwhile to elaborate two further parameters for the graphical representation of experimental avoidance data. These parameters were: (a) the crossing time (T) of the avoidance values, and (b) the fluctuation (σ_e) around the avoidance value. Both were derived from the computer outputs of the time series analysis and were measured as the standard deviation of the avoidance value. The crossing time T defined the reaction time which elapsed until the fish first achieved the avoidance value for the particular experiment. The fluctuation (σ_e) was measured as the standard deviation of the avoidance value. When T was considered in conjunction with the σ_e, supplementary information about the experiment was produced, by plotting time T versus the time spent by the fish in clear water as a percentage of the total time. This new representation is called herein 'avoidance reaction representation' for each experiment.

The industrial effluent used was taken from the effluent of a representative metal plating industry. Toxic substances in the effluent are chromium (20 mg/l), nickel (50 mg/l) and small amounts of iron (4.0 mg/l). The 96 h LC_{50} of the effluent was found to be an effluent–water mixture containing 60% effluent after analysis of three effluent samples at the Environment Canada laboratories of Longueuil, Quebec. Other properties of the effluent measured are: pH, 6.8, total solids, 540 ppm; and grease and oils (<2.5 ppm).

In evaluating synergistic and antagonistic phenomena between the actual industrial effluent and its components, synergism was defined as the joint action of two or more toxicants injected simultaneously and antagonism as the opposing action (Brown, 1968).

A very simple approach which has been used in classifying the interactions of two toxicants is to expose fish to half the concentration of toxicant A necessary to produce a given response and half the concentration of toxicant B necessary for the same response. If this combination first causes the response, the actions of A and B are exactly additive. If it causes more than the given response, they are more than additive. If the combination does not produce the response, the toxicants are either less than additive, show no interaction or are antagonistic (Sprague, 1970).

For the prediction of avoidance reactions the same basic idea of synergistic and antagonistic interactions between the constituent pollutants of the effluent were applied, using proportionally various concentrations of the three predominant toxicant components of the industrial effluent, and comparing them with the avoidance reaction to the, actual industrial effluent itself.

Finally, three parameters were used for the evaluation of the safe toxic levels. These were: the maximum allowable toxicant concentration (MATC), the safe concentration (SC) and the application factor (AF).

The distinction between SC and MATC exists because what is allowed is not always safe.

F is a fractional factor applied to acute toxicity tests to estimate the toxicant concentration that may be safe for chronic or lifetime exposure of test organisms. An application factor is unitless and is calculated as the ratio of MATC to the 96-hour LC_{50} or incipient lethal level (ILL). Sometimes SC is used instead of MATC.

$$AF = \frac{\text{MATC}}{\text{ILL or 96-h } LC_{50}} \quad \text{or} \quad AF = \frac{\text{SC}}{\text{ILL or 96 h } LC_{50}}$$

For each toxicant there is obviously a numerical value for the ratio of the SC to the acutely lethal concentration. The use of the 'safe-to-lethal' ratio is often an expedient suggesting that a level of protection will be offered without necessary experimental work. Twenty-one tests were conducted to determine the effect of different concentrations of the actual effluent on the avoidance behavior and the distribution of the fish (rainbow trout) in the channel. Another twenty-one tests with the three main components of the previous effluent (Cr, Ni, Fe) were conducted in order to ascertain the contribution of the individual metals to the avoidance behavior of the fish exposed to the raw effluent at concentrations equivalent to ones that were found in the chemical analysis.

Three times per experiment, 500 ml of water from the polluted side of the channel was removed through four siphons and analyses were made for Cr, Ni or Fe, using atomic absorption after extraction to detectable limits. The limit of detection was 2 μg/l and the precision ± 1 μg/l. Results of the tests are given as a function of the LC_{50} concentration and the actual concentration can be obtained by dividing by 0.60. For example, a concentration of 1 : 250 LC_{50} yields an actual effluent concentration of 1 : 417 (dilution ratio).

8.3 RESULTS AND DISCUSSION

The longitudinal distribution of fish in the industrial effluent experiments shown in Fig. 8.2 indicated that there was a massive shift of fish from sections 3 and 4 to section 2, at effluent concentration of 1 : 200 LC_{50}. At the same time there was a change from the polluted side to the non-polluted side of the channel. The distribution appeared to change dramatically as the effluent concentration was increased. With the higher concentrations there was a tendency for the fish to move into the clean water behind the separator in Section 2 of the channel. This behavior occurred because Section 2 was one of the locations where the fish could avoid the effluent.

Fig. 8.3, 8.4 and 8.5 showed the longitudinal distribution of fish in the Cr, Ni, and Fe experiments independently. In particular, Figs. 8.3 and 8.4 indicated the trend for the fish to cluster in the second section of the channel. This effect could explain the marked concentration of fish in the second section observed also during the testing of the actual effluent (Fig. 8.2). Due to the low level of Fe present in the actual effluent in comparison with the levels of Ni and Cr, the effect of the concentration of fish in the third section in the Fe experiments (Fig. 8.5) had little impact on the actual effluent distribution (Fig. 8.2).

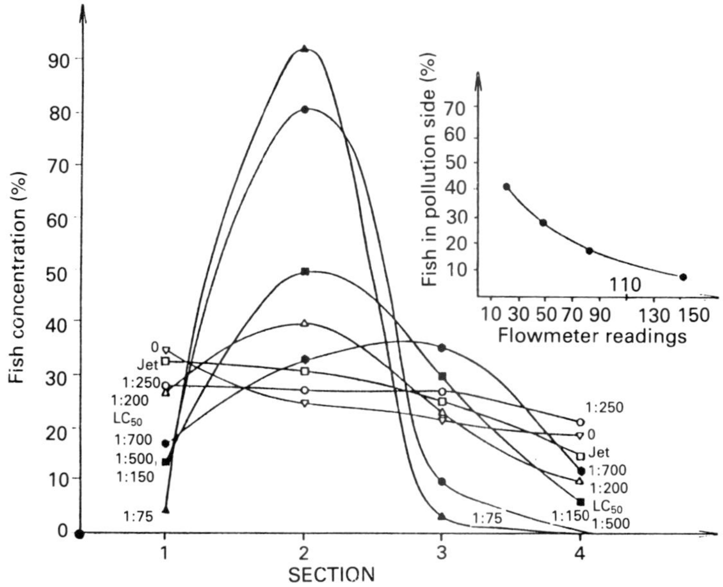

Fig. 8.2 — Longitudinal distribution of fish in industrial effluent experiments.

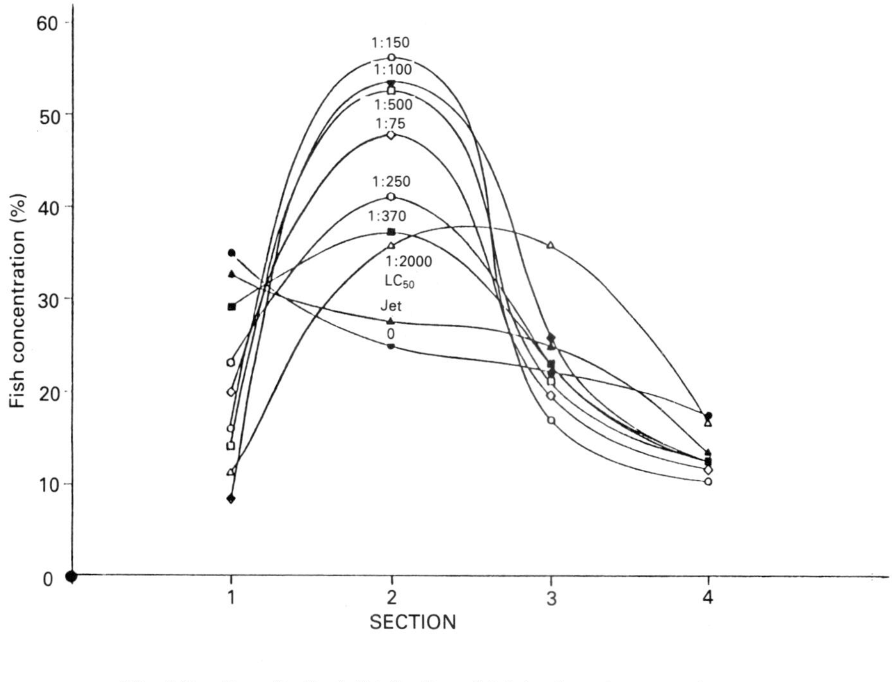

Fig. 8.3 — Longitudinal distribution of fish in chromium experiments.

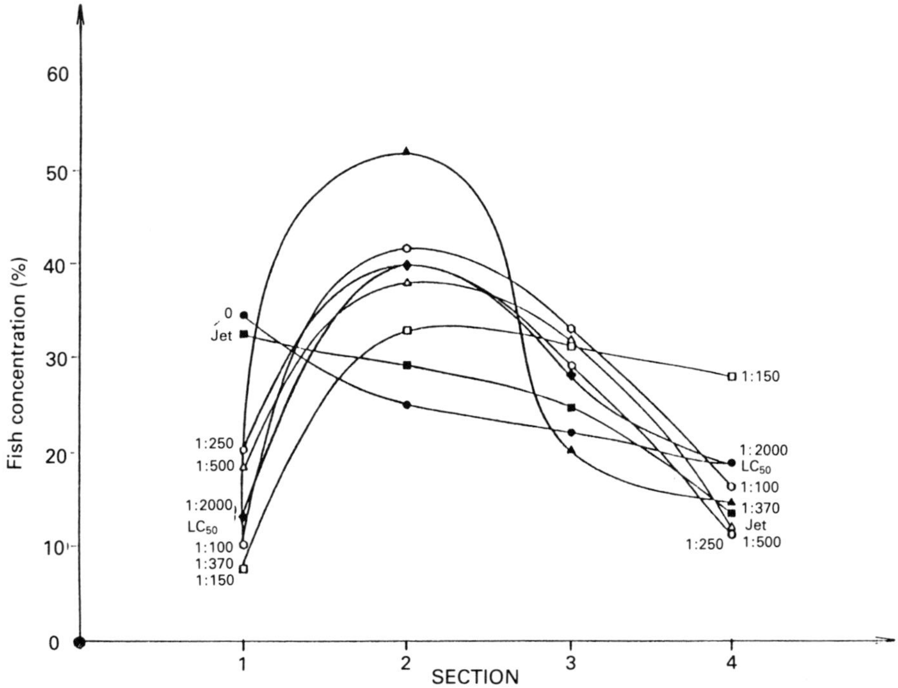

Fig. 8.4 — Longitudinal distribution of fish in nickel experiments.

The ANOVA results for the longitudinal distribution of the actual effluent, Cr and Ni indicated that in Sections 1, 3 and 4 the percentages of fish in clear water for the different concentrations, were not significantly different (i.e. they belonged to the same group, A). In Section 2 which was significantly different from the other three sections, using Duncan's tests, two groups, A and B, were found, with the percentage of fish at concentrations 1:75, 1:100 and 1:150 LC_{50} being different from the percentages in the remaining concentrations.

The reference points in each of Figs. 8.2, 8.3, 8.4 and 8.5 for the evaluation of the effect of the different pollutants to the longitudinal distribution throughout the channel were: (a) the 0 concentration, indicating longitudinal distribution of the fish in clean water, and (b) the jet distribution which takes into account the distribution of the fish in relation to a clean water jet.

The vertical distribution of fish in the industrial effluent experiments shown in Fig. 8.6 did not change significantly, indicating a movement of the fish to a similar elevation in the non-polluted side.

Figs. 8.7, 8.8 and 8.9 showed the vertical distributions of fish in the experiments with Cr, Ni and Fe respectively. In Fig. 8.7, a trend was observed for fish to move from depth A (bottom) to depth C (surface) with increasing concentrations of Cr, in contrast to Fig. 8.8 which showed an increasing preference for depth A and a decreasing preference for depth C with increasing concentration of Ni. The contradictory reaction of the vertical distribution of fish to Cr and Ni led to the observed

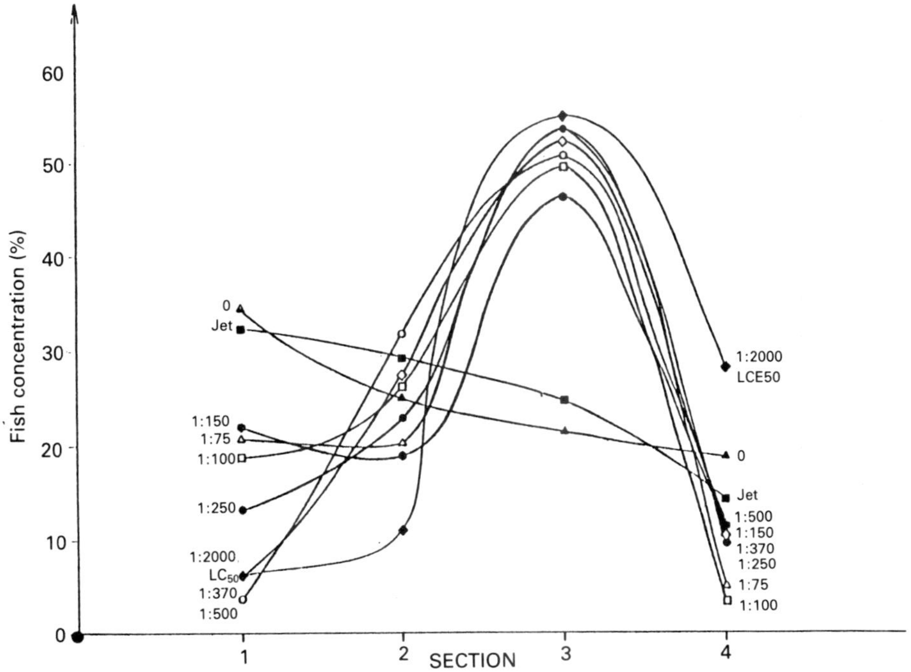

Fig. 8.5 — Longitudinal distribution of fish in iron experiments.

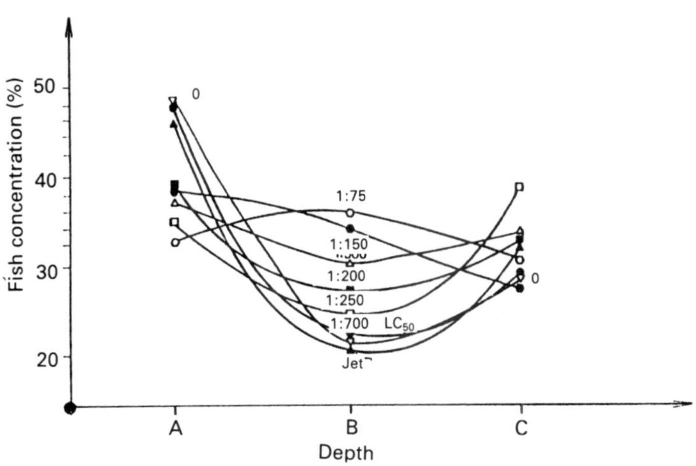

Fig. 8.6 — vertical distribution of fish in industrial effluent experiments.

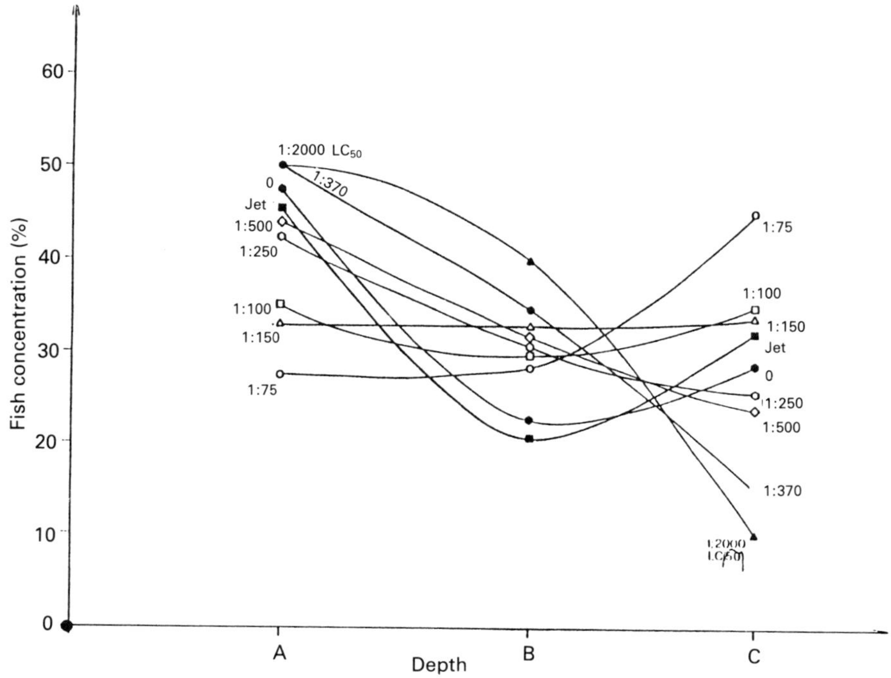

Fig. 8.7 — Vertical distribution of fish in chromium experiments.

considerably smaller difference between the three depths of the vertical distribution of fish in the industrial effluent experiments (Fig. 8.6). Fig. 8.9 showed a small fluctuation around the even distribution, presumably because of the very small amounts of injected Fe. The ANOVA results indicate that there was a significant difference between the three depths for all the vertical distributions (actual effluent, Ni, Cr, Fe), whereas the results of Duncan's tests indicated no significant differences between the various concentrations at each depth. The test results shown in Fig. 8.6 tend to indicate that whereas a difference in the vertical distribution appeared between the three depths, it was unclear that this was only due to the effluent. Part of the problem was the limited depth of the channel.

Referring to Fig. 8.10, it can be seen that avoidance of the industrial effluent began at very low concentrations and was already evident at a value of 1 : 700 LC_{50}. Extrapolation of the avoidance curve would indicate that avoidance can possibly begin as early as 1 : 3000 LC_{50}. The final avoidance was of the order of 90%, indicating that some of the fish did not avoid the effluent all the time even at high concentrations. This could conceivably be explained by impairment of sensory organs in unresponsive fish.

Figs 8.11, 8.12 and 8.13 represented the avoidance curves for Cr, Ni and Fe respectively.

In Fig. 8.11 the Cr avoidance curve showed that an increase in the Cr concentration from 0.015 mg/l to 0.444 mg/l (equivalent to the Cr present in the actual effluent) did not have significant effect. The avoidance reactions for these levels of Cr remained between 46–54% (Fig. 8.11).

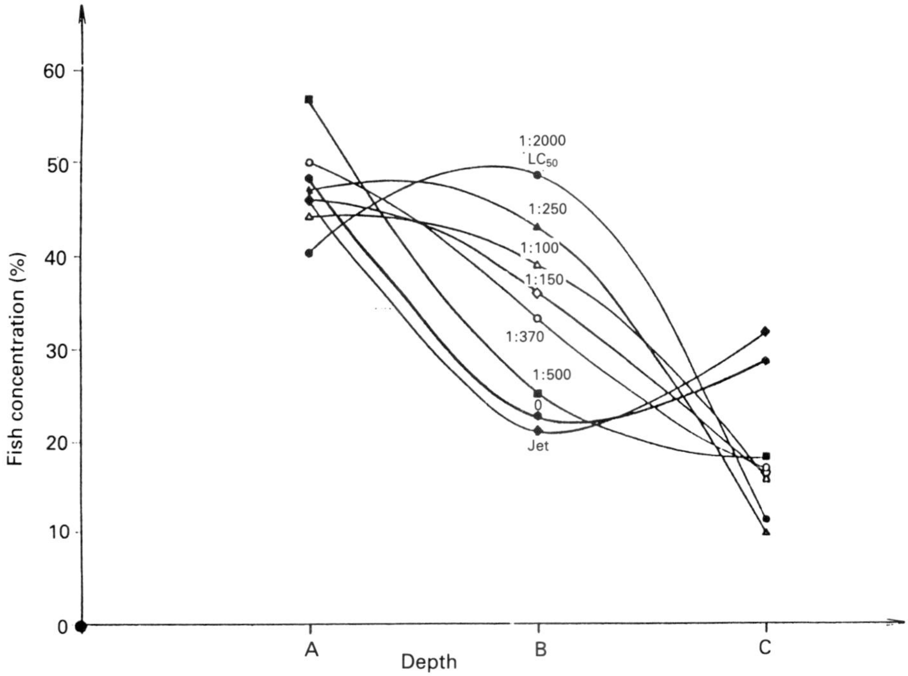

Fig. 8.8 — Vertical distribution of fish in nickel experiments.

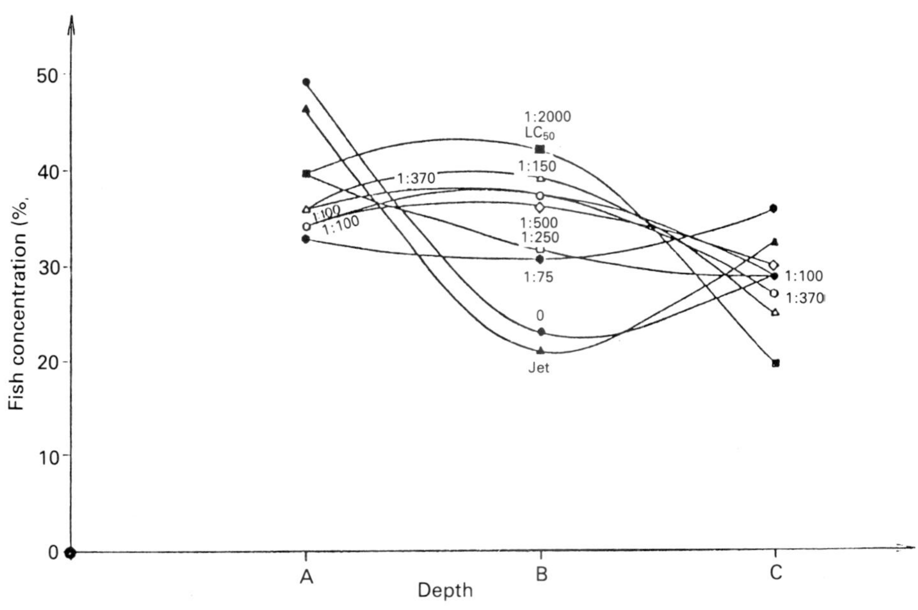

Fig. 8.9 — Vertical diustribution of fish in iron experiments.

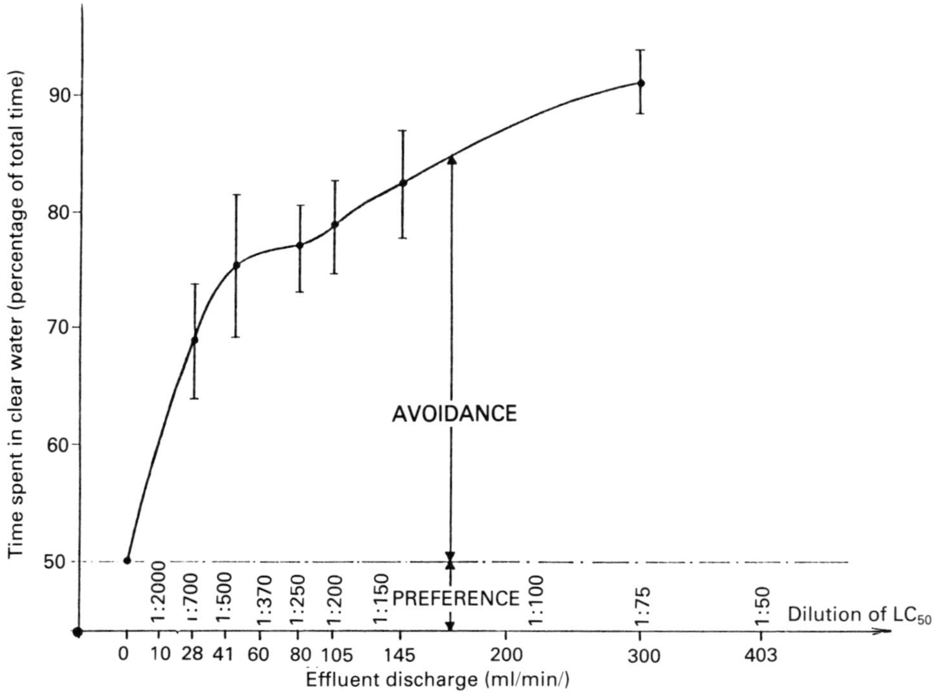

Fig. 8.10 — Avoidance curve of plating industrial effluent.

The avoidance reaction of Fe (Fig. 8.13) remained stable between 50 and 55% because of the low level of injected Fe concentrations.

An increasing avoidance reaction with increasing Ni concentrations was apparent (Fig. 8.12). It was concluded that Ni was the main contributor to the avoidance reaction of fish exposed to the actual effluent. At a concentration of 1 : 100 LC_{50} avoidance reaction of Ni was 90%.

Fig. 8.14 presents the avoidance reaction representation (T, σ_e, overall avoidance, dilution ratio of LC_{50}) for the experiments of the actual industrial effluent conducted. It appeared that T fluctuates in a range between five and 30 minutes (80% between 15 and 25 minutes) and the σ_e between 4 and 11% (80% between 5 and 10%).

Figs. 8.15–8.17 present the avoidance reaction representation for the Cr, Ni and Fe experiments.

In Figs. 8.15–8.17, T fluctuated in a range of 8–30, 7–35 and 8–32 respectively. Considering all cases, the maximum time to achieve stability of the responses on time was 35 minutes.

The implication of the data portrayed in Figs. 8.14–8.17 is that avoidance reaction representation provides supplementary experimental avoidance information, by evaluating the reaction time which elapsed before the avoidance value for a particular experiment was first achieved, and also the variability of the avoidance value for every experiment.

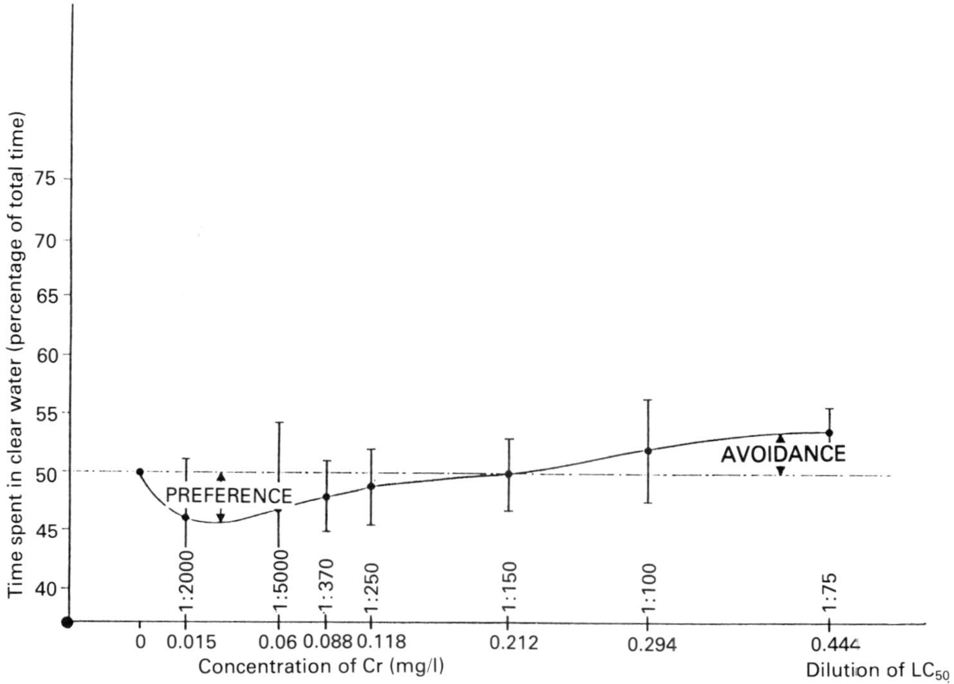

Fig. 8.11 — Avoidance curve of chromium.

Each one of the basic components of the industrial effluent (Cr, Ni, Fe) contributed a certain proportion to the overall avoidance. Fig. 8.18 illustrates the avoidance of the industrial effluent and its components. The evaluation of synerginistic and antagonistic phenomena was derived for each concentration injected. From Table 8.1, in order to interpret the synerginistic and antagonistic phenomena for each concentration, one compares the avoidance of the industrial effluent (column 2) with: (a) the dominant avoidance reacxtion to Ni (column 4), (b) the proportional additive avoidance of the three components, Cr, Ni, Fe (column 6), and (c) the proportional additive avoidance of the two dominant components, Cr, Ni (column 7).

Table 8.1 (column 8) illustrates that, at very low concentrations (e.g. 1:2000 LC_{50}) the avoidance response of Cr, Ni and Fe was exactly additive. At concentrations between 1:700 LC_{50} and 1:150 LC_{50}, synergism among the three heavy metals was observed in a decreasing range from more than additive to less than additive. Finally, at higher concentrations than 1:100 LC_{50}, antagonistic reactions were observed, implying that the presence of Fe and Cr in the effluent contributed to the observations that the final effluent avoidance was less than the avoidance to the Ni alone.

In general, it appeared that Ni was the dominant component of the industrial effluent and was the main contributor to the avoidance reaction. Cr played a secondary role and the presence of Fe was almost insignificant.

Experiments using a full spectrum of concentrations of the main components of

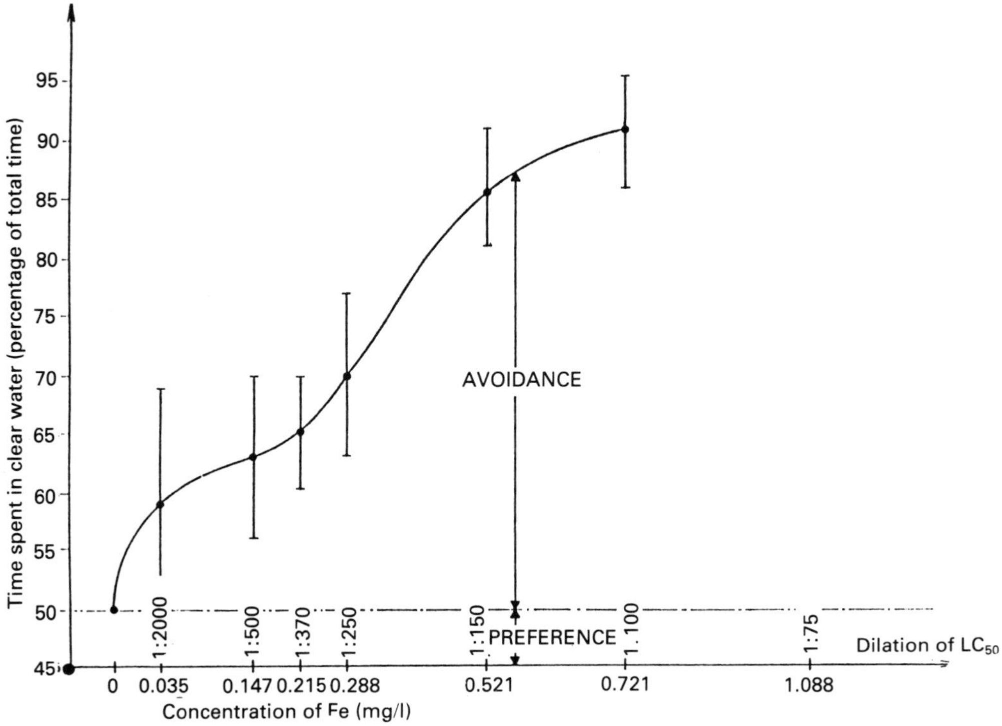

Fig. 8.12 — Avoidance curve of nickel.

the industrial effluent (from very low ones to lethal ones), provides the information for an evaluation of the synergistic and antagonistic phenomena of the main components, while allowing speculation about possible significant effects of other elements present in the effluent at very low or trace concentrations.

Finally, Table 8.2 lists the pollutants used in column 1 and the range of the 96-hour acute lethal LC_{50} in column 2. The MATCs, the SCs and the AFs from the most recent published references (Clarke, 1974; EPS 1973; Cherry, 1982; Sauter *et al.*, 1976, Hale, 1977; Pickering, 1980; Cairns and Dickson, 1978; Becker and Wolford, 1980) are presented in columns 3, 4 and 5 respectively. Columns 6 and 7 contained values for avoidance threshold and avoidance reaction derived from Figs 8.10–8.13 (0.2 mg/l for total Cr, 0.01 mg/l for Ni, 0.01 mg/l for Fe and 1 : 3000 LC_{50} for the actual plating effluent). Finally, column 8 gives proposed adjustments of the safe level concentrations for the pollutants studied based on literature data and the results of this study. These adjustments include the MATC, SC and AF for each pollutant, and effluent limits for the principal chemicals of the actual industrial effluent, having considered the concept that, in general, safety is related to the 'no-effect' (below thresholds) sublethal concentrations.

8.4 CONCLUSIONS

(1) An avoidance apparatus has been designed and implemented, using advantages of previous avoidance systems and at the same time providing new advantages.

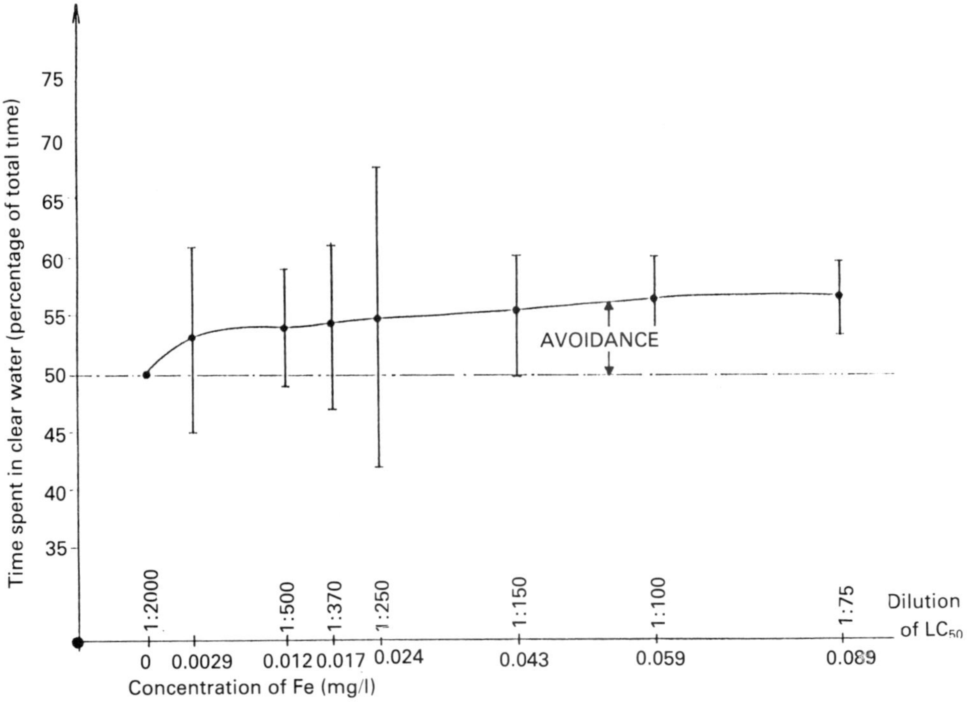

Fig. 8.13 — Avoidance curve of iron.

The innovative characteristics of the design were the capacity for vertical, longitudinal and lateral analysis of the position of test organisms; the use of a large number of test organisms; and the ability to test different pollutants, to store results for later re-examination, and to use advanced video and analysis techniques for data interpretation.

(2) The spectrum of plating-industry effluent concentrations injected ranged between1 : 3000 and 1 : 75 LC_{50} (well below the lethal threshold) and produced an avoidance reaction between 50 and 90%. This illustrated a significant potential hazard to aquatic organisms at sublethal concentrations which would be impossible to predict or estimate with the conventional static LC_{50}.

(3) The sublethal threshold concentration for the avoidance reaction to the plating industry effluent was 1 : 3000 LC_{50}. The avoidance reaction reached 90% in dilution ratio of 1 : 75 LC_{50}.

(4) The longitudinal and the vertical distribution of the tested organism's position, were mainly related to the preference–avoidance pattern of the polluted and non-polluted sections of the channel and partly related to the effluent characteristics.

(5) Nickel was the dominant component of the industrial effluent used and the main contributor to the avoidance reaction. Cr played the secondary role and the presence of Fe was insignificant.

(6) The comparison of the avoidance reaction to the industrial effluent: (a) with Ni

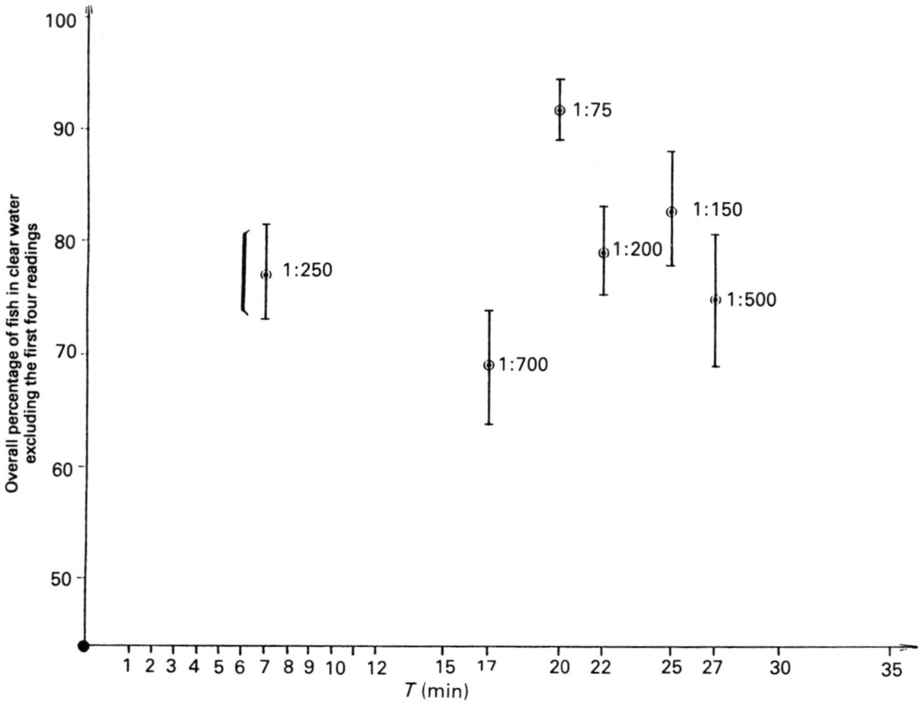

Fig. 8.14 — Avoidance reaction representation of fish in industrial effluent experiments

(b) with the additive avoidance of Ni and Cr, and (c) with the additive avoidance of Ni and Cr and Fe, led to the estimation of the synergism and antagonism zones for the three heavy metals with the variation of the actual effluent's injected concentration. In very low concentrations, effluent effects were exactly additive, from 1 : 700 to 1 : 150 LC_{50} effects ranged from more than additive to less than additive and in higher concentrations effects were antagonistic.

(7) The avoidance reaction representation was proposed as a new manner of depicting supplementary experimental avoidance information related to the stability of the reactions with time.

(8) Combining information from literature and findings from this study, adjustments of the safe level concentrations for the pollutants studied were proposed, including values for maximum allowable toxicant concentration, safe concentrations and application factors, and also effluent limits for the actual plating effluent.

ACKNOWLEDGEMENTS

The assistance of Dr Gilles LaRoche, Terry Rigby, Norman Birmingham, Christian Blaise, Dr Robert Hutcheon, Joan Cornell and Lesley-Ann Judge is gratefully appreciated.

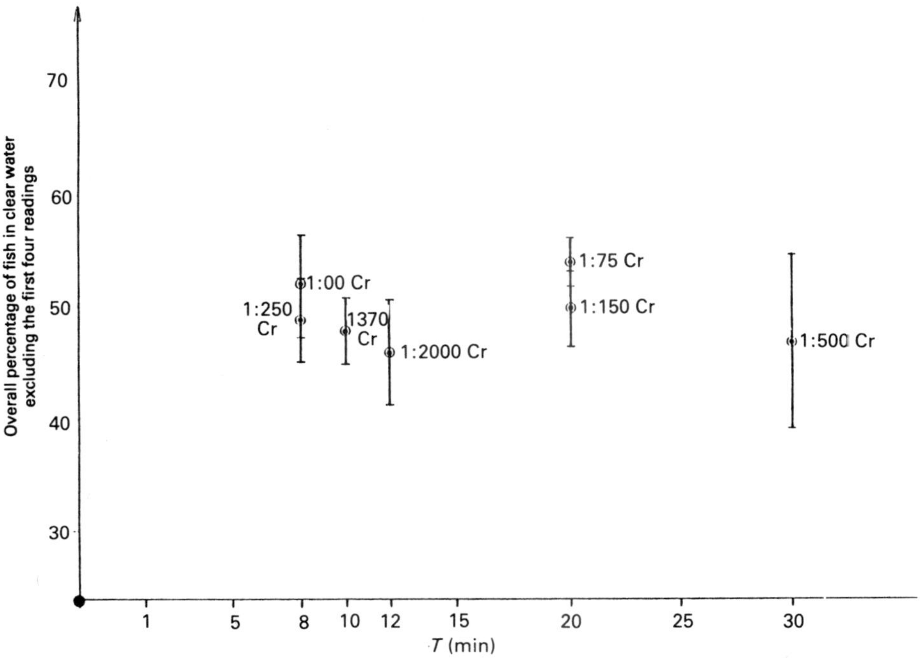

Fig. 8.15 — Avoidance reaction representation of fish in chromium experiments.

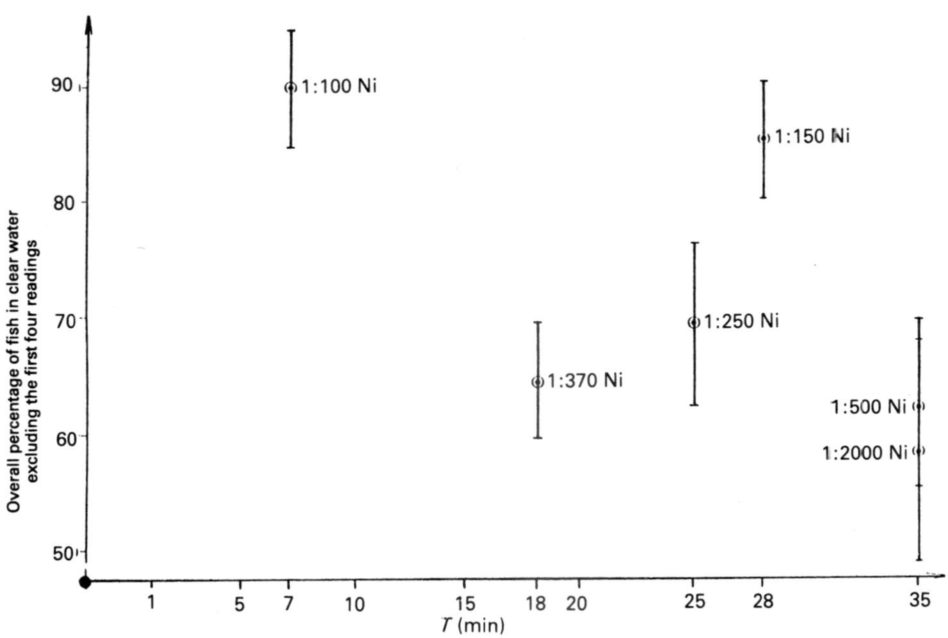

Fig. 8.16 — Avoidance reaction representation of fish in nickel experiments.

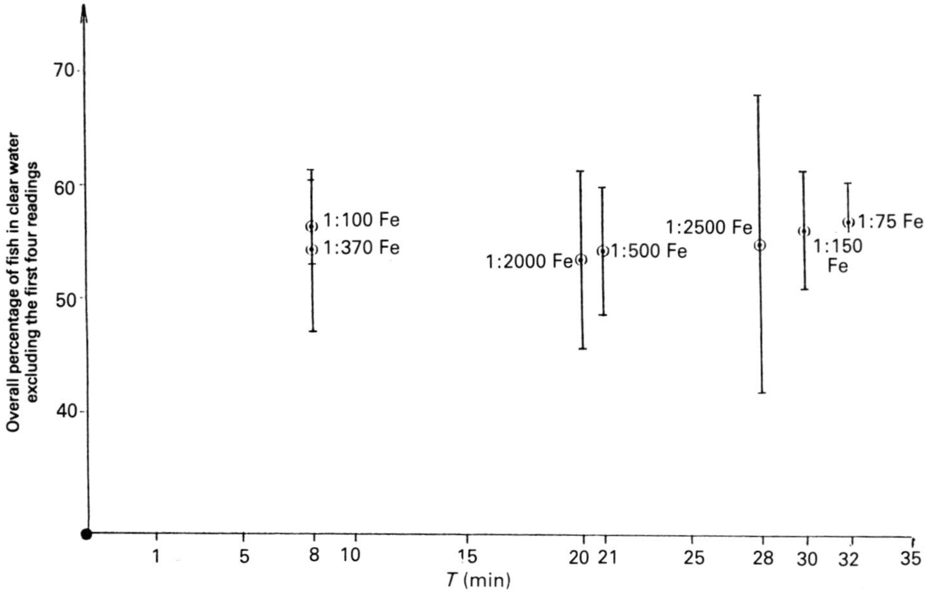

Fig. 8.17 — Avoidance reaction representation of fish in iron experiments.

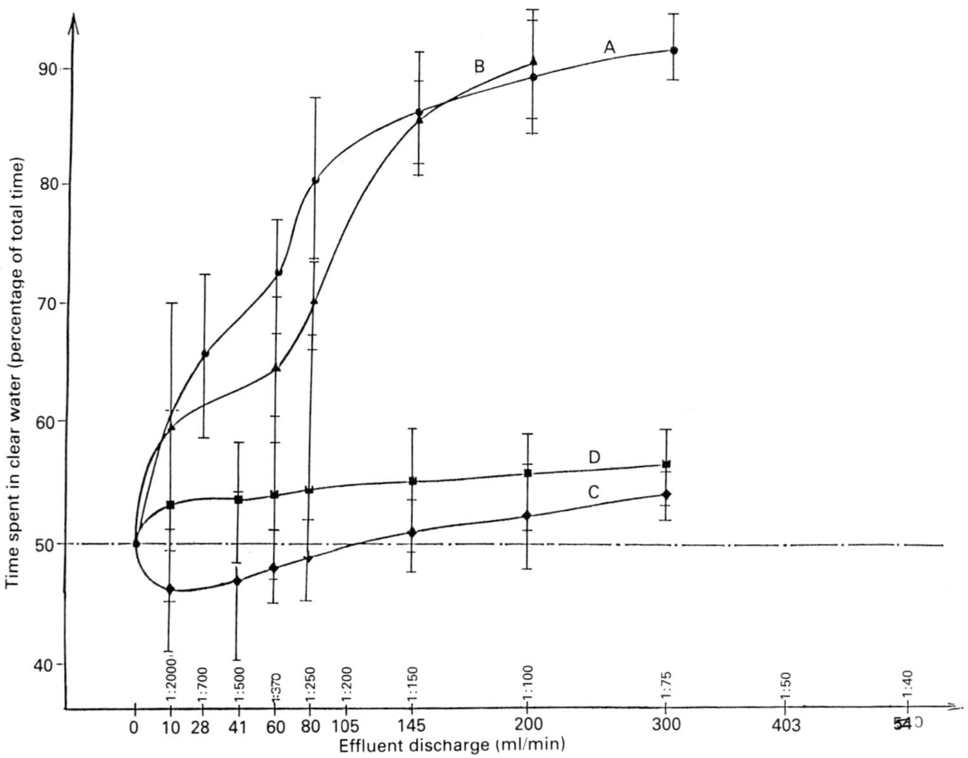

Fig. 8.18 — Avoidance of industrial effluent and its components. A, industrial effluent; B, nickel; C, chromium; D, iron. Ni:Cr:Fe=12.5:5:1.

Table 8.1 — Synergism and antagonism in the avoidance reaction of the industrial effluent and its components

Effluent dilution of LC_{50}	Avoidance of industrial effluent (%)	Proportional avoidance of Cr (%)	Proportional avoidance of Ni (%)	Proportional avoidance of Fe (%)	Additive avoidance of Ni, Cr, Fe (%)	Additive avoidance of Ni and Cr (%)	Remarks about synergism and antagonism
(1)	(2)	(3)	(4)	(5)	(6)= Proportional addition of (3)+(4)+(5)	(7)= Proportional addition of (3)+(4)	(8)=Comparison between (2) and (4), (6) and (7)
1 : 2000	57.5	46.0	59	53.0	58	62.0	Exactly additive
1 : 700	65.0	46.5	61	53	60.5	64.0	Synergism More than additive
1 : 500	68.70	47.0	64	53.5	64.5	67.5	Synergism More than additive
1 : 370	72.0	48.0	65	54.0	67	69.0	Synergism More than additive
1 : 250	80.6	49.0	70	54.0	73	74.0	Synergism More than additive
1 : 200	83.5	50.0	75	54.5	79.5	79.5	Synergism More than additive
1 : 150	85.8	51.0	85	55.0	91.0	90.0	Synergism Less than additive
1 : 100	88.0	52.0	90	56.0	98.0	96	Antagonism
1 : 75	90.5	54.0	94	56.0	>100	>100	Antagonism

Synergism Anatgonism

Table 8.2 — Proposed adjustments to the safe toxic level for the pollutants of this study for rainbow trout (*Salmo gairdneri*, Rich)

Pollutant	96-h LC_{50} in (mg/l)	MATC in (mg/l)	Safe concentration in (mg/l)	Application factor	Avoidance threshold in (mg/l)	Avoidance reaction	Proposed adjustments
(1)	(2)	(3)	(4)	(5)	(6)	(7)	(8)
Chromium (hexavalent	69[a]	0.050[a,f]	0.025[c]	0.03	—	—	MATC = 0.05 SC = 0.03 AF = 0.03
Chromium (trivalent)	100[b,g]	0.10[e]	0.10[a]	0.04	—	—	MATC = 0.12 SC = 0.1 AF = 0.04
Chromium (total)	17–118[b,e]	0.050[a]	0.050[a]	0.03	0.2	0.5 mg/l 55%	MATC = 0.050 SC = 0.025 AF = 0.03
Nickel	20–40[a,d,b]	0.04[a,b]	0.025[a]	0.01	0.01	1 mg/l 95%	MATC = 0.02 SC = 0.01 AF = 0.005–0.008
Iron	1–2[b]	1.0[a]	0.5[a]	0.25	0.001	0.1 mg/l 55%	MATC = 0.02 SC = 0.01 AF = 0.1
Actual plating effluent	60%	Dilution ratio 1 : 1380	Dilution ratio 1 : 2760	—	1 : 3000 LC_{50}	1 : 40 LC_{50} 93%	Effluent limits Cr(tot)=1.0 mg/l Ni=1.5 mg/l Fe=2 mg/l

[a]Clarke (1974); [b]EPS (1973;[c]Cherry and Cairns (1982); [d]Hale (1977); [e]Sauter (1976); [f]Pickering (1980); [g]Cairns and Dickson (1978); [g]Becker and Wolford (1980).

REFERENCES

Becker, C. D. and Wolford, W. G. (1980) Thermal resistance of juvenile salmonids sublethally exposed to nickel, determined by the critical thermal maximum method. *Environm. Pollution* **21A** 181.

Birtwell, I. K. (1977) A field technique for studying the avoidance of fish to pollutants. *Proceedings of the 3td Aquatic Toxicity Workshop, Halifax, N.S.*, Technical Report No. EPS-5-AR-77-1, pp. 69–86. Environmental Protection Service, Halifax, Canada.

Brown, V. M. (1968) The calculation of the acute toxicity of mixtures of poisons to rainbow trout. *Water Research* **2** 723–733.

Cairns Jr, J. and Dickson, K. L. (1978) Field and laboratory protocols for evaluating the effects of chemical substances on aquatic life. *J. Test. Eval.* **6** 81–90.

Cherry, D. S. and Cairns Jr, J. (1982) Biological monitoring. Part V — Preference and avoidance studies, *Water Research* **16** 263–301.

Clarke, R. McV. (1974) *A summary of toxicity information for major effluent components from inorganic chemical industries*. Department of Environmental Fish and Marine Services, Technical Report Series No. CEN/T-74-9 Aquatic Toxic Studies Division — Res. Manag. Branch, Winnipeg.

EPS (1973) *Water quality criteria*, EPS-R3-73-0 3, (commonly referred to as 'Blue Book'). Environmental Protection Service, Halifax.

Hadjinicolaou, J. and Spraggs, L. D. (1981) Methodology for assessing toxicity avoidance. Proceedings of the 8th Aquatic Toxicity Workshop in Guelph, Canada. *Can Tech. Rep. of Fish. Aquat. Sci.* pp. 68–82.

Hadjinicolaou, J. and Spraggs, L. D. (1982) Toxicity avoidance of toxic effluents, Proceedings of the 9th Aquatic Toxicity Workshop in Edmonton, Alberta. *Can. Tech. Rep. Fish. Aquat. Sci.* No. 1163, pp. 210.

Hale, J. G. (1977) Toxicity of metal mining wastes. *Bulletin Environm. Contam. Toxicology* **17** 66.

Höglund, L. B. (1951) A new method of studying the reactions of fishes in stable gradients of chemicals and other agents, *Oikos* **3** 246–267.

Ishio, J. (1969) Behaviour of fish to exposed substances, In: O. Tang (ed.) Advances in water pollutioon research, Vol. 1, Proc. 2nd Int. Conf. held in Tokyo, Pergamon Press, New York pp. 19–40.

Jones, J. R. E. (1947) The reactions of *Pygosteus pugitius* to toxic solutions. *J. Exp. Biol.* **24** 11–122.

Kleerekoper, H. and Morgensen, J. (1963) Role of olfaction in the orientation of *Petromyzon mainus*. I. Response to a single aurine in prey's body odour. *Physiol. Zool.* **36** 347–360.

Kleerekoper, H. (1967) Some aspects of olfaction in fishes with special reference to orientation. *Am-Zool.* **7** 385–395.

Pickering, Q. H. (1980) Chronic toxicity of hexavalent chromium to the fathead minnow. *Arch. Envir. Contam. Toxicol.* **9** 405.

Sauter, S. *et al.* (1976) Effects of exposure to heavy metals in selected fresh water fish: toxicity of copper, cadmium, chromium and lead to eggs and fry of seven fish species. EPS-600/3-76-105, U.S. EPA, Duluth, Minnesota.

Scherer, E. and Nowak, J. (1973) Apparatus for recording avoidance movements of fish. *J. Fis. Res. Board Can.* **30** 1594–1596.

Shelford, V. E. and Allee, W. C. (1913) The reactions of fishes to gradient of dissolved atmospheric gases. *J. Exp. Zool.* **14** 107–166.

Spraggs, L. D., LaRoche, G. and Hadjinicolaou, J. (1981) *Integrated toxicity avoidance study — final report.* Report for the Canadian Department of Supply and Services.

Spraggs, L. D., Gehr, R. and Hadjinicolaou, J. (1982) Polyelectrolyte toxicity tests by fish avoidance studies. *Water Science and Technology* **14** 1564–1567.

Sprague, J. B. (1964) Avoidance of copper–zinc solutions by young salmon in the laboratory. *J. Water Pollut. Cont. Fed.* **36**(8) 990–1004.

Sprague, J. B. (1970) Measurement of pollutant toxicity to fish II — Utilizing and applying bioassay results. *Water Research* **4** 3–32.

Wells, M. M. (1915) Reactions and resistance of fishes in their natural environment to acidity, alkalinity and neutrality. *Bio. Bull.* **29** 221–257.

Westlake, G. F. and Lubinski, K. S. (1976) A chamber to monitor locomotor behaviour of free swimming aquatic organisms exposed to simulated spills. *Proceedings of the national conference on control of hazardous materials spills, New Orleans, LA.*

9

Developing portable computer-automated biomonitoring for a regional water quality surveillance network

Eric L. Morgan
Department of Biology and Center for the Management, Utilization and Protection of Water Resources, Tennessee Technological University, Cookeville, Tennessee 38505, USA

Richard C. Young,
President, Young-Morgan & Associates, Inc., Brentwood, Tennessee 37207, USA

James R. Wright Jr,
Division of Air and Water Resources, US Tennessee Valley Authority, Knoxville, Tennessee 37901, USA

9.1 INTRODUCTION

A water-quality management plan employing automated aquatic biological monitoring systems (automated biomonitoring) was first proposed two decades ago by Cairns (1967) and Cairns *et al.* (1970 a,b). They envisioned the implementation of automated biomonitoring systems at critical wastewater discharges along a watershed drainage as a quality control technique in the management of aquatic ecosystems. This approach would have the advantage of providing an early warning to developing toxicity associated with specific discharges and in the upstream dilution or reference waters. To accomplish this, fish gill-ventilatory (breathing) rates and swimming activities would be continuously monitored, responses processed for stress events by computer and results made available to users on a real-time basis. By viewing these results with those from routine in-stream biological surveys and physical water-quality samplings in the light of a resource management plan for the watershed, a water-quality management network was conceived. If implemented, both effluent and in-stream water sources would be monitored continuously and toxic episodes validated through the routine stream-survey (biological monitoring) assessments (Cairns *et al.*, 1972, 1973). In effect, a complementary monitoring program had been proposed for the present-day approach to water quality-based toxicant-control advocated by the US Environmental Protection Agency (US EPA, 1985).

In 1977, within a decade after the automated biomonitoring watershed-management plan had originally been proposed and with the advantages of solid-state electronics becoming readily available, remote applications for a regional water-quality monitoring network were being developed (Morgan *et al.*, 1977, 1980). These portable, computer-automated biomonitoring devices were data-linked with earth-satellite communication options. Moreover, it was proposed that a regional watershed-monitoring network approach incorporate portable biosensing devices not only at remote locations, but in mobile bioassay laboratories and buoy configurations (Fig. 9.1).

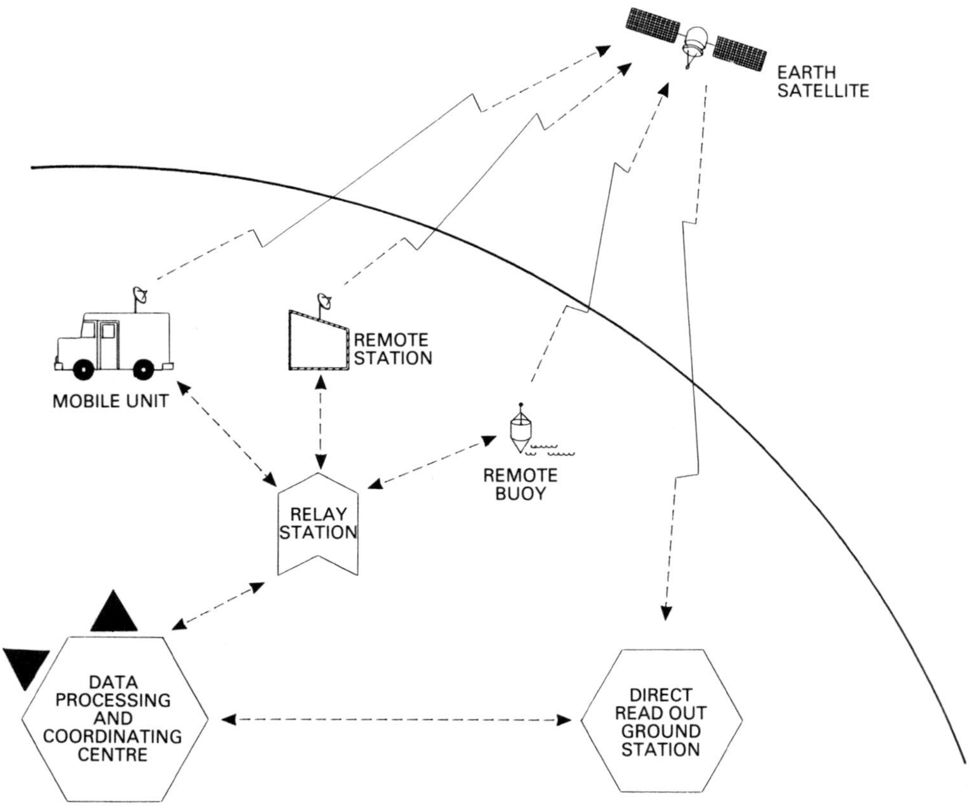

Fig. 9.1 — Proposed regional network using portable aquatic automated biomonitoring systems.

Our efforts have involved an ongoing series of laboratory and field trials to develop and test portable aquatic biosensing devices applicable to a real-time network system. Our objective here is to provide a review of specific technical developments crucial to the proposed network. In this regard, emphasis will be placed on reviewing recent applications of portable computer-automated biomonitoring systems developed for mobile bioassay laboratories; remote *in situ* implementations at satellite data-linked platforms; streamside research laboratory testing; and proposed future operations.

9.2 BACKGROUND

Appropriate designs and various configurations were considered before testing prototype systems under simulated applications. In the process of developing automated biosensing detectors that would have the ability to be interfaced with remote water quality data-collection platforms (DCPs) yet operate from mobile bioassay laboratories and streamside facilities, biosensing detectors were operated under field conditions. Biological concerns were also addressed, including chamber size, feeding, and longevity of test subjects. Following these screening tests, suitable configurations were selected and are now being employed *in situ* at water-quality platforms on remote streams in the Southern Appalachian Mountains (Morgan *et al.*, 1986), in a mobile bioassay laboratory (Morgan *et al.*, 1988), and from a streamside laboratory (Morgan and Young, 1984). These ongoing research projects constitute key elements in the proposed regional network supported by portable automated biomonitoring devices.

9.3 REMOTE BIOSENSING PLATFORMS

Prototype automated biosensing devices were operated at remote water quality stations on two drainages between 1976 and 1981. One trial utilized a warm-water station on the Cumberland River, near Carthage, Tennessee and a second trial was implemented on a small cold-water stream in the Great Smokey Mountains National Park. The different types of fish-holding chambers, DCPs and physical monitors employed in these initial studies are discussed in detail elsewhere (Morgan *et al.*, 1980, Morgan and Eagleson, 1981, 1983). Benefiting from the experiences of these trials, new amplifier designs and a general-purpose computer-assisted DCP were selected for the ongoing studies incorporating these applications.

The six rainbow trout chambers that are presently being used at each of the remote biosensing stations now in operation were constructed from 10×30 cm plastic pipe (Morgan *et al.*, 1986). Each chamber houses an individual free-swimming fish. Chambers were positioned horizontally in-stream where exposure to ambient flow-through conditions was continuous. Two stainless steel probes extending horizontally the length of the chamber were connected to separate co-axial cables. The two cables were attached to individual amplifiers allowing the detection of each opercular voltage differential of a trout placed between the probes inside the chamber. The six low-noise variable-gain, d.c.-powered differential amplifiers constructed for each station were mounted in a weatherproof instrumentation box positioned streamside. The box also included registers which accumulated breathing events from each fish prior to being submitted to the d.c.-powered, computer-assisted DCP. Thus, bioelectric action potentials induced by discrete breathing events by individual fish were amplified to give a representative analogue voltage of adequate amplitude. In this study, analogue frequencies were converted to digital pulses and stored in registers prior to processing by the microprocessor controller of the DCP. The specific advantages and general operational characteristics of the amplifier and the DCPs selected for these studies has been presented in more detail in an earlier article (Morgan *et al.*, 1986).

Signals generated by individually monitored trout were accumulated for a 15-minute interval each half-hour. The data were held in the memory of the programmed DCP and transmitted to the National Oceanographic and Atmospheric Administration, Geostationary Operational Environmental Satellite System (NOAA, GOES) on six occasions each day. Broadcast data received by satellite were retransmitted to a direct-readout ground station (DRGS) at the US Tennessee Valley Authority (TVA), Data Services Branch, Knoxville, Tennessee (Fig. 9.2).

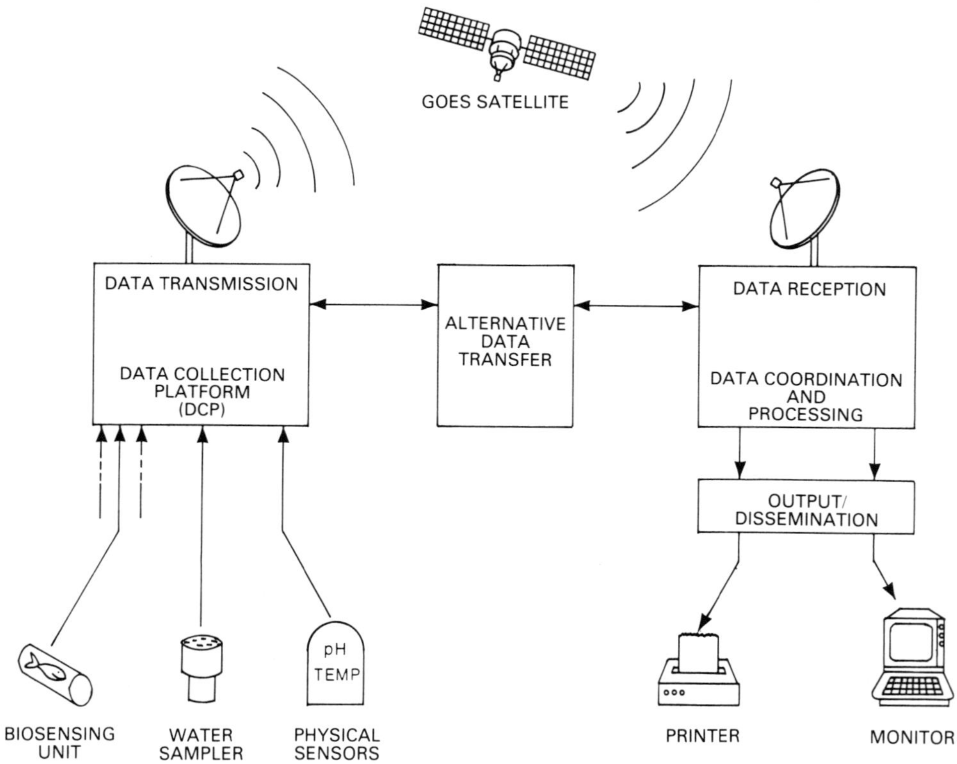

Fig. 9.2 — Remote water-quality monitoring network.

Simultaneous transmission of water-quality parameters for this particular application included: temperature, hydrogen ion concentration (pH), specific conductance, and stream-stage elevation. Physical sensors were positioned in-stream adjacent to trout-holding chambers and instrumentation housed in small prefabricated structures located above the flume. The DCP was programed to initiate a time-sequenced water-sampler during critical water-quality events or by fish breathing-rate changes.

9.3.1 Field trials from remote platforms

In the winter of 1987, two satellite data-linked remote water-quality DCPs were placed in full operation along a 4.3-km reach of Laurel Branch Creek, a tributary to the North River drainage of the Tellico River watershed in the Southern Appalachian Mountain of Tennessee. Both DCPs were equipped with the six biosensing devices designed to record trout breathing responses and the physical sensors as previously described. These remote water-quality platforms, along with a recently added third platform, are being maintained as a part of a five-year Acid Precipitation Mitigation Program. This program was initiated in 1985 by the US Fish and Wildlife Service as a part of the National Acid Precipitation Assessment Program to evaluate stream ecological responses to regulated liming practices.

A third DCP has been positioned in the head waters of a contiguous drainage, the South Fork Citico Creek, which is located in the Cherokee Wilderness Area. Though the same type platform and biosensing units have been used, several modifications were necessary due to wilderness status mandates. Government regulations precluded the construction of facilities and prohibited the use of service vehicles in the area; thus eliminating the flume for flow monitoring and the antennae construction necessary for satellite communications. Therefore, instead of the satellite transceiver which was used at the other two platforms, a magnetic tape cassette device for data recording has been substituted. Cassettes and the two 12 V d.c. batteries used at the DCP are serviced during monthly visits when water and biological samples are collected from this reach of the stream. Data from magnetic tapes are managed and processed in a similar manner to satellite retrieval data.

9.3.2 Data retrieval and intepretation

Serially transmitted data recieved by the data coordinating computer at the DRGS were converted into a spreadsheet format using a customized dBASEIII plus program†. the resulting ASCII files were then imported into a LOTUS 1-2-3 program‡ for data manipulation, preliminary statistical analysis and graphing. Typically, fish breathing rates were normalized by averaging the individual, discrete events recorded for each fish to produce average breathing rates for the specified time intervals. Statistical estimates of sample data variance were obtained for each normalized group. Graphing the fish breathing rate over time permitted visual observation of periods of increased or decreased activity. These data were then compared to water quality data collected simultaneously from real-time sensors or concomitant analyses of water samples to evaluate the case and/or validity of the stress event, in this application acid precipitation episodes (Morgan *et al.*, 1986).

† Ashton-Tate, Inc.
‡ Lotus Development Corporation.

Several statistical programs were available for more detailed analyses (SAS Institute, 1979) though a graded response procedure may best be measured by a modification of the non-parametric method of the Mann–Whitney U-test (Snedecor and Cochran, 1980).

9.4 MOBILE LABORATORY-BASED PORTABLE AUTOMATED BIOMONITORING

The facilities and equipment used in this study consisted of a mobile bioassay laboratory equipped with automated monitoring systems and a portable environmental isolation locker. The locker was used for housing the animals to be monitored (Fig. 9.3). The mobile bioassay laboratory was positioned streamside on the tailwater below Douglas Dam on the French Broad River (FBR) at km 39, in East Tennessee. A 45-day trial was completed during the spring the spring of 1986 (Morgan *et al.*, 1988). Adjacent to the trailer and 5 m from the river, the portable styrofoam-insulated environmental locker was positioned to receive water from a gravity-supply head tanks. Using two water-supply systems, one was set up to deliver recirculating–replenishment reference water, and the other administered once-through water pumped from the river.

Inside the locker, two groups of eight, 38-l styrofoam aquaria with covers were set up to receive pass-through flows of approximately 0.5 l/min from the gravity-feed head tanks. Water supply systems for both groups of eight aquaria were designed so that flows could be mixed or delivered singly as reference and/or river water. On the inside surface of the styrofoam covers, a low-intensity yellow–red LED (light-emitting diode) lamp provided a 12/12-h photoperiod. Positioned inside each aquarium was a fish chamber similar in design to those used in the remote biosensing studies. A total of 16 sauger weighing about 200–450 g were initially maintained under reference water flows for a nine-day acclimation period.

Inside the trailer, sauger breathing events were received via co-axial cables, amplified 5×10^5 times before being digitized for temporary storage in the interface register-board located in the desk-top computer. Differential amplifiers were identical to those used in the remote biosensing work except an LED lamp and power supply was included on each board. This provided a visual clue to each breathing event. Using the automated fish respiration monitoring system (AFIRMS)-Basic program, data could be requested at an interval of 1 to 59 minutes as selected by the operator (Morgan *et al.*, 1988, Morgan and Young, 1984). Here a four-minute cycle was used, i.e. data were recorded for a three-minute interval followed by a one-minute wait period. Data were logged in an ASCII format and filed on hard disk. Accompanying the breathing data and taken once every four-minute cycle were physical sensor measurements for temperature, pH, dissolved oxygen and specific conductance. A similar though non-portable system without the real-time advantages of a portable computer was developed by Cairns and Gruber (1979) and Cairns *et al.* (1980).

9.4.1 Field trials from the mobile laboratory

After a nine-day acclimation period to reference water flows, one group of eight sauger was subjected to FBR water whereas the remaining group of eight continued

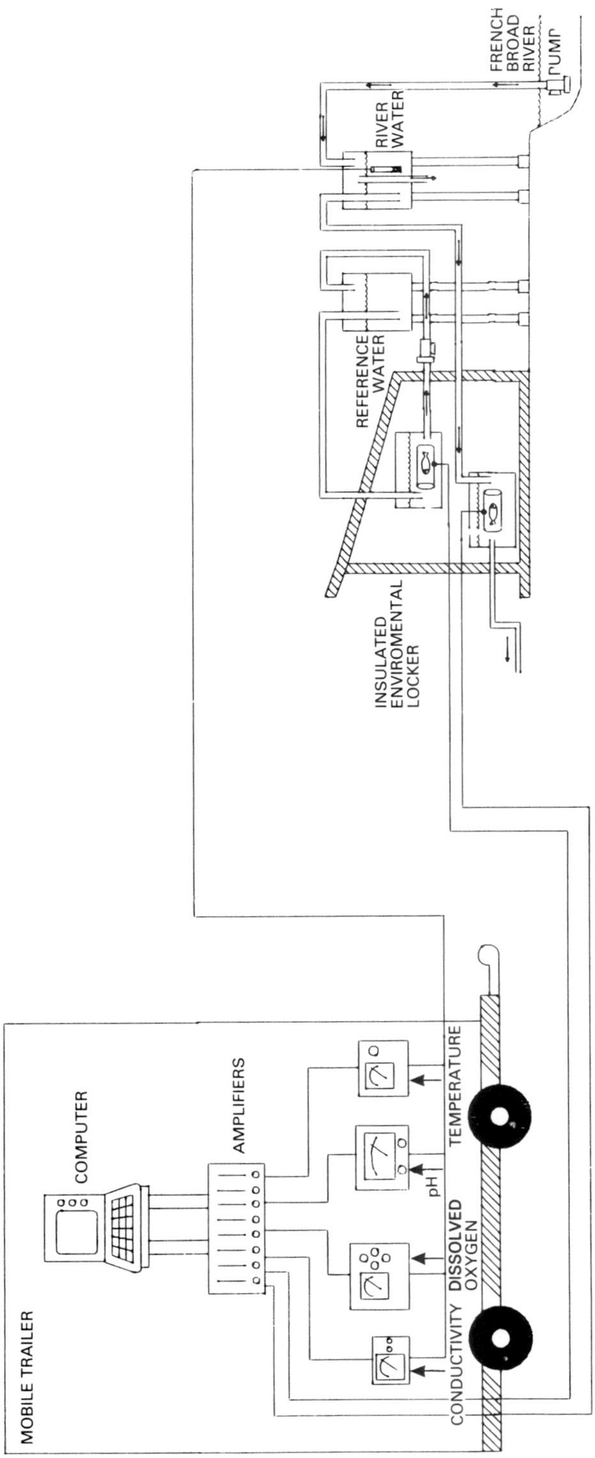

Fig. 9.3 — Schematic of portable automated aquatic biomonitoring system operating from mobile bioassay laboratory at streamside.

to receive reference flows. About one-half the 1.3 m^3 volume of recirculating reference water supply was replenished daily. Over the remaining 35 days of the study, when sauger breathing rates were continuously recorded, complementary aquatic toxicity tests were performed in the laboratory with water samples taken at five locations along the FBR reach below Douglas Dam. Reference water taken from the nearby Little Region River was used as dilution water. Fathead minnow larval survival and growth studies and *Ceriodaphnia* spp survival and reproduction success tests (Horning and Weber, 1985) were done at the US TVA, Aquatic Research Laboratory (Morgan *et al.*, 1988).

Following more than 40 days of field testing with the AFIRMS operating continuously under ambient river-water conditions, 12 sauger were subjected to a stepwise treatment over a four-hour period of approximately 0.9 and 1.9% toxic leachate collected from a landfill. In previous tests, this leachate was found to cause significant mortality at less than 1% leachate concentration to *Ceriodaphnia* and fathead minnows in static-renewal short-term chronic laboratory toxicity tests.

9.4.1.1 Data management and system evaluation

Utilizing the AFIRMS-Basic program to manage input data and to establish files, breathing rates were stored as events per minute. Similar to the data management scheme for remote sensing information, these ASCII files were imported into a LOTUS 1-2-3 program for preliminary evaluations. Should more intensive statistical tests be indicated, files could be processed in one or more of the same statistical programs and procedures used for remote sensing data.

Field trials utilizing the portable systems from streamside and the toxicant evaluations done in the laboratory using *Ceriodaphnia* and fathead minnows revealed little or no toxicity in FBR waters below Douglas Dam during the spring of 1986. These results may not reveal a true picture of ambient river toxicity since stream flows were unusually low due to a record drought for this period of the year. No major point source industrial or municipal discharges occurred between the test site and the dam some 10 km upstream. Should toxic substances be present in runoff flows following storm events in the drainage, then this potential contribution to the toxic evaluation was not assessed. At streamside, when approximately 1.9% leachate was delivered to 12 sauger being monitored by the AFIRMS, immediate increases in breathing rates resulted. Within several hours of exposure mortalities had occurred and after 13 hours, 11 of the 12 sauger displayed no breathing responses and were found to be dead upon examination. Aside from problems associated with algae-clogged water-supply systems, the automated system performed reasonably well as a real-time portable biomonitor and results support future application in mobile laboratories.

9.5 STREAMSIDE LABORATORY USING MULTIPLE SPECIES

Field trials for this aspect of portable automated biomonitoring were done in a streamside research laboratory operated by the US TVA (Morgan and Young, 1984). Techniques and procedures for monitoring bioelectric signals produced by various physiological functions from the mussel and mayfly nymph subjects were developed, incorporating modified methods previously tested in computer-automated fish biomonitoring (Morgan and Young, 1984). The portable computer, interface units, and amplifiers used in these trials were similar to those employed in the mobile bioassay laboratory studies, except that preliminary testing was done with two prototype AFIRM systems. These earlier experimental systems differed in having the AFIRM-Basic program executed from EPROM instead of a hard-disk sector and were equipped with 12 instead of 24 channels.

The observation that certain aquatic animals induced rhythmic analogue bioelectric frequencies associated with specific physiological functions, i.e. heart rates, breathing events, and other muscular action-potential responses, was central to automation (Camougis, 1960; Cairns, 1970). A more detailed description of the actual field trials using multispecies applications had previously been reported (Morgan and Young, 1984). To summarize holding chambers used for bluegill sunfish were modifications of those used in remote biosensing studies. Typically this cylindrical chamber houses a single free-swimming fish and a pair of probe-type antennae, arranged on the cylinder in such a way that frequencies generated by each gill ventilatory response could be readily recorded by the accompanying portable computer. Mayfly nymph chambers were designed along a similar plan. For example, using artificial tube chambers consisting of Tygon or plastic tubing which had been equipped with micro-probe antennae were occupied by burrowing mayfly nymphs. Frequencies produced by abdominal gill oscillations within the chamber could readily be measured by electrophysiology methods (Fig. 9.4) and the automated biomonitor. Finally, micro-probes were attached between the valves at the anterior and posterior regions of freshwater mussels. In addition to gill movement patterns and foot adductor muscle events, bioelectric action potentials induced by each peristaltic contraction of the heart were monitored (Morgan and Young, 1984).

A multispecies complement was realized by combining a bluegill chamber, a mussel chamber, and a burrowing mayfly nymph chamber into one rectangular glass tank (38 l). A total of 16 amplifiers were used to make up a group of four multispecies complements. Discreet responses from each animal were monitored as events per minute and data were managed by the computer using the same AFIRMS-Basic program for automated fish respiration rate monitoring. Baseline results reveal that many different types of aquatic animals may serve as biosensors and that continuing research is needed to further identify specific bioelectric events that reflect functional measures of behavioral and physiological stress.

9.6 FUTURE DEVELOPMENTS

Portable automated biomonitoring studies completed over the past decade have provided the experience needed to show that these configurations are not only inexpensive, but can be operated from mobile facilities and remote streamside stations. The recent decision by the US TVA to install satellite communications at its DCP-equipped regional stage-height and rain-gauge stations provides a future opportunity for cooperation in developing a regional drainage-basin network which could include automated multiple biosensing devices. Additionally, since other agencies such as the US Geological Survey and the US Army Corps of Engineers employ similar DCP-equipped water-quality stations, future options for inclusion of automated biomonitoring exist.

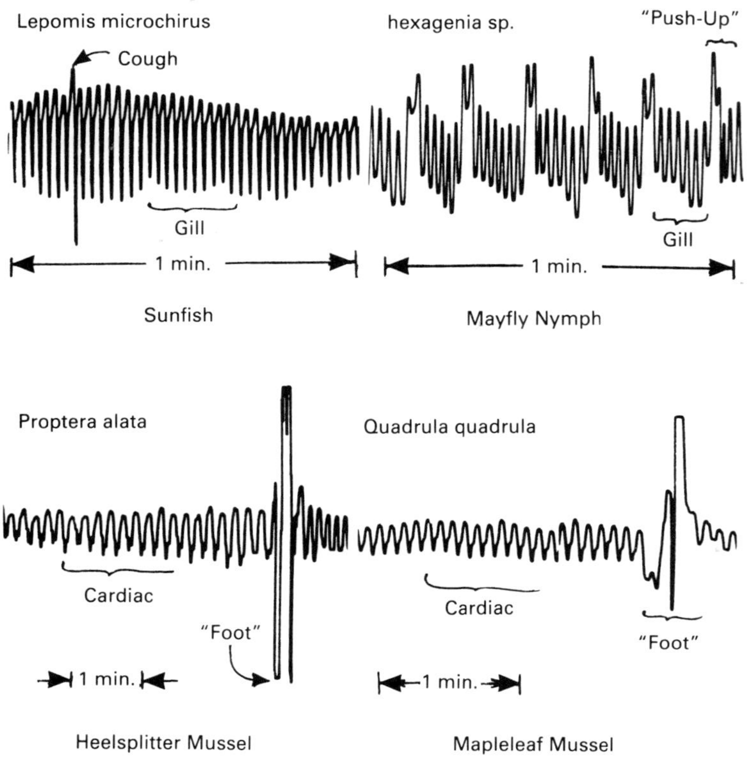

Fig. 9.4 — Representative physiological responses monitored under ambient water-quality exposures by automated multispecies biosensor at remote streamside station, Center Hill Lake, Tennessee.

For possible future applications, a prototype biosensing buoy device was constructed and tested in the laboratory (Morgan and Eagleson, 1981). Buoy components consisted of aquatic animal-holding chambers attached to a waterproof flotation compartment fabricated from PVC (polyvinyl chloride) plastic. Housed inside the compartment was the electrical components consisting of a d.c. power supply, amplifiers, minicomputer and transmitter (Fig. 9.5). Fish opercular activity was received by the minicomputer in digital form before conversion to a radio frequency (RF) for broadcasting by the transmitter. In this particular application, RFs transmitted were representative of equivalent respiratory frequencies. The RF data received on shore were intercepted by a similar portable minicomputer which in turn processed the information for storage to magnetic tape and for display. The buoy configuration has only received proof-positive testing and awaits more intensive development before applied to permanent applications. For example, fish tagged with transmitters could be released in a drainage system partially contaminated with toxic or known mutagens or potential carcinogens. Frequencies transmitted from released fish would be received by scanning receivers housed in fixed buoys. Various data communication linkages could be used, including earth-satellite

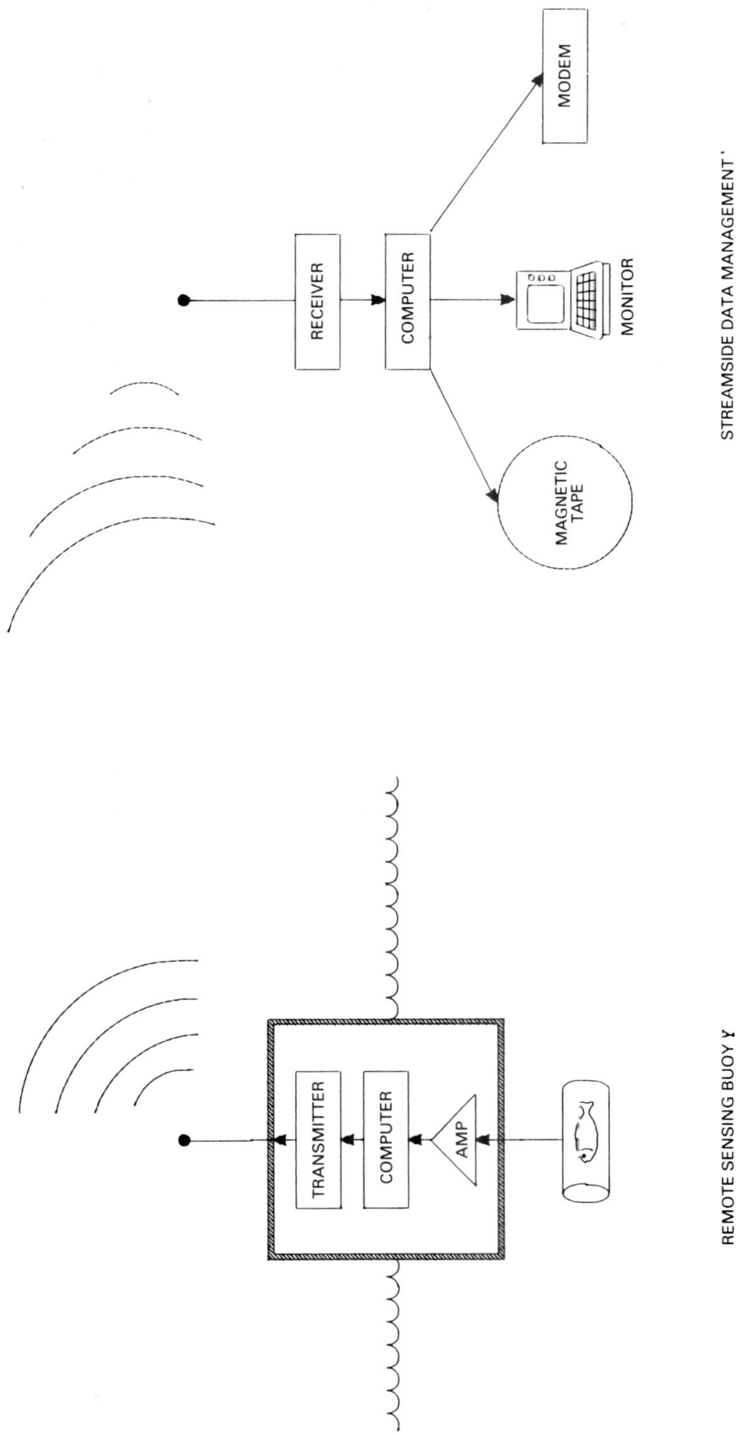

Fig. 9.5 — Remote biosensing buoy application.

retrieval from remote areas. Each tagged fish occupying the monitored area would be identified, its residence time recorded and relative location specified. Should a substance that tends to bioaccumulate be of interest, then individual animals could be selectively sampled for analysis. The advantages of a buoy-monitored approach are potentially important to risk-assessment monitoring enabling mitigative efforts at hazardous waste clean-up sites. Validation data needed in meeting specified contaminant limits may best be acquired in this manner. Such information would be useful in real-time situations and in simulation models for dynamic bioaccumulation of toxic substances by fish gill-exchange, i.e. the US EPA-GETS (gill exchange for toxic substances) model (Suarez *et al.*, 1987).

Presently, instead of relying on the rather crude measure of breathing rate as the dependent variable for judging stress, we are using the inherent information found in the analog frequencies recorded for breathing events. The theoretical analog form used here is shown in Fig. 9.6. By applying current biomedical statistics procedures

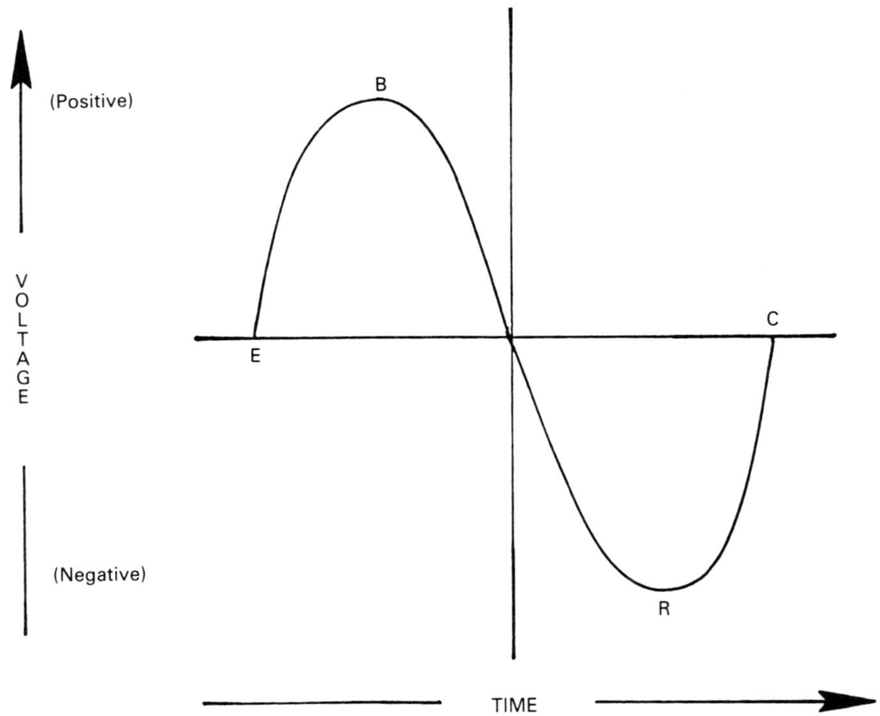

Fig. 9.6 — Typical analogue waveform representing one complete opercular ventilation or breathing event, E and C, zero intercept; B and R, peak positive and negative amplitudes.

routinely used in electrophysiological analyses, we hope to identify stressed breathing frequencies. Information evaluated relative to stressed breathing events includes shifts in positive and negative slope functions, peak amplitudes, areas or volumes produced and time relationships. The use of Fourier transformation methods of discrete frequencies composing the gill ventilatory analog signal is presently underway and is believed to be a more sensitive measure of stress than that being employed by Evans and Johnson (1985). Should this approach prove useful, it may have merit not only in network monitoring, but possibly in future systems proposed for space-station water-supply management and earthquake early warnings (Morgan *et al.*, 1987).

Finally, research that may lead to future applications includes a fish respirometer interfaced to the automated fish breathing-rate biomonitor (Fig. 9.7). Employing a

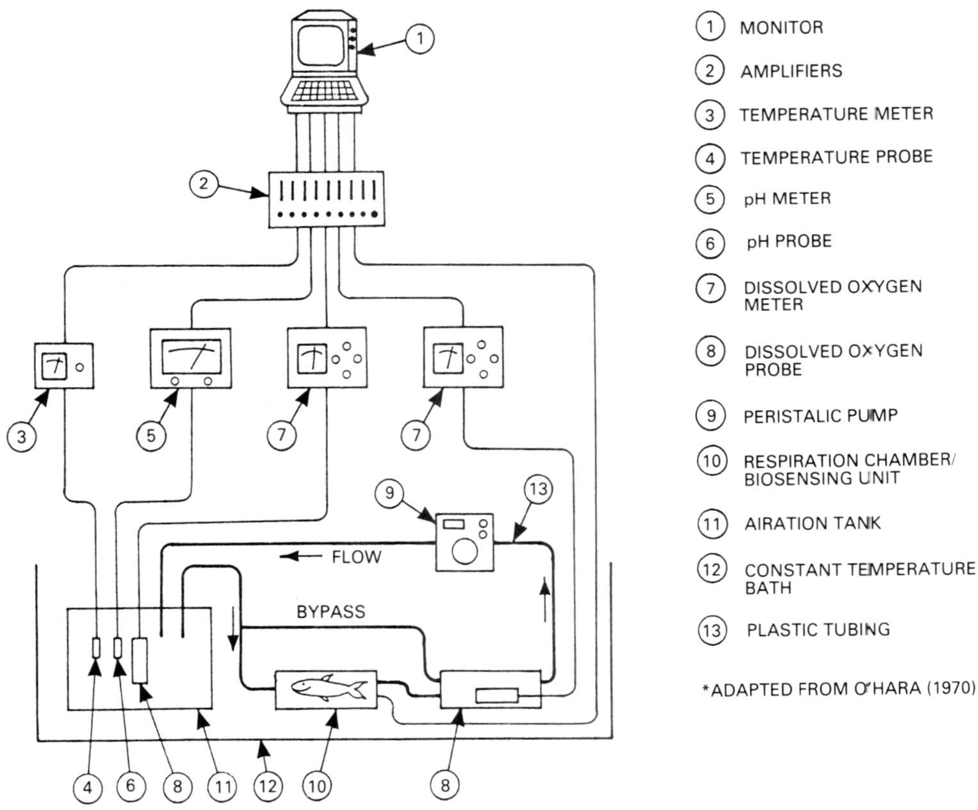

Fig. 9.7 — Computer-monitored fish respiration chamber. 1, monitor; 2, amplifiers; 3, temperature meter; 4, temperature probe; 5, pH meter; 6, pH probe; 7, dissolved-oxygen meter; 8, dissolved-oxygen probe; 9, peristaltic pump; 10, respiration chamber/biosensing unit; 11, aeration tank; 12, constant-temperature bath; 13, plastic tubing. Adapted from O'Hara (1970).

plastic tube chamber in the laboratory, a fish respirometer was set-up to detect an oxygen concentration differential of a constant flow between the influent and effluent waters. Coupled to the respirometer chamber was a pair of antennae used to receive action potentials from each breathing event. Both dissolved-oxygen concentrations and breathing events were continuously logged on the accompanying portable computer every other minute. From these data a relationship can be developed when a mass balance approach is taken where oxygen (in mg) is plotted against breathing events. The slope of the relationship between oxygen utilized and breathing activity may be viewed in the light of the basal metabolism of the aquatic animal of given size at a specific temperature. Preliminary studies reveal that by changing these and other extrinsic factors, or by subjecting the animal to a substance that alters intrinsic metabolism, a concomitant shift in the slope of relationship occurs.

ACKNOWLEDGEMENTS

Ongoing support for remote application is made available through US Department of Interior, US Fish and Wildlife Service; Eastern Energy Land Use Team, via Tennessee Department of Health and Environment; and the Center for Management, Utilization, and Protection of Water Resources, Tennessee Technological University. Support for the mobile bioassay system is provided by US Tennessee Valley Authority, Division of Air and Water Resources with technical support through Data Services Branch. We gratefully acknowledge the Tennessee Wildlife Resources Agency and the US Department of Agriculture, US Forest Service, Cherokee National Forest for making facilities available and for field assistance.

Support for the developmental phase of this project was provided by funds made available to the Tennessee Water Resources Research Center by the following participants: Environmental Science Division; Oak Ridge National Laboratory; Office of Natural Resources; US Tennessee Valley Authority; and Office of Water Research and Technology, US Department of the Interior (USDI). Additional funding was provided by the US Corps of Engineers, Nashville District, and Aquatic Ecology Fund, Environmental Biology Research Program, Tennessee Technological University.

REFERENCES

Cairns Jr, J. (1967) Use of quality control techniques in the management of aquatic ecosystem. *Wat. Resourc. Bull.* **3**(4) 47–53.

Cairns Jr, J. (1970) *Jour. Water Pollu. Control Fed.* **42**(1) 77–82.

Cairns Jr, J. *et al.* (1970) *J. Wat. Pollut. Control Fed.* **42** 685–703.

Cairns Jr, J. *et al.* (1970) Biology Systems as Pollution Monitors, Research/ Development. **9**(70) 22–24.

Cairns Jr, J., Sparks, R. E. and Waller, W. T. (1972) The design of a continuous flow biological early warning system for industrial use, *Purdue Univ. Engng. Bull.* **141** 242–255.

Cairns Jr, J., Hall, J. W., Morgan, E. L., Sparks, R. E., Waller, W. T. and Westlake, G. F. (1975) *The development of an automated biological monitoring system for water quality.* Bulletin 59. Virginia Water Resources Research Center, VPI&SU.

Cairns Jr, J. and Gruber, D. (1979) *BioScience* **29**(11) 665–669.

Cairns Jr, J. *et al.* (1980) *Water Res. Bull.*, **16**(3) 421–427.

Camougis, G. (1960) Recording bioelectric potentials from aquatic animals. *Turtox News* **38** 156–167.

Evans, G. P. and Johnson, D. (1985) Automated detection of pollution of drinking water intakes using fish. *Wat. Supply* **3** (Erratum) 179–186.

Horning II, W. B. and Weber, (1985) *Short-term methods for estimating the chronic toxicity of effluents and receiving water to freshwater organics.* US EPA/600/4-85/014.

Morgan, E. L. *et al.* (1977) Biological water quality monitoring from remote stations and N.O.A.A.-GOES satellite. *Proceedings of the 4th Joint Conference on Sensing of Environmental Pollution,* New Orleans, November 1977, pp. 885–887.

Morgan, E. L. *et al.* (1980) *Water International,* **5** 23–27.

Morgan, E. L. and Eagleson, K. W. (1981) *Automated biomonitoring applications to remote water quality surveillance and time rated toxicity assay.* Technical report No. 56, p. 51. Tennessee Water Resources Research Center, The University of Tennessee, Knoxville, Tennessee.

Morgan, E. L. and Eagleson, K. W. (1983) Remote biosensing employing fish as real-time monitors of water quality events. *In Proceedings of an International Symposium on Hydrological Applications of Remote Sensing and Remote Data Transmission,* Publication No. 145. International Association of Hydrological Science.

Morgan, E. L. and Young, R. C. (1984) *Verh. International Verin. Limnology,* **22**, 1432–1435.

Morgan, E. L. *et al.* (1986) Linking automated biomonitoring to remote computer platforms with satellite data retrieval in acidified streams. In B. G. Isom, J. Bates and S. Dennis (eds) *Impacts of Acid Rain and Deposition on Aquatic Biological Systems,* STP 928, American Society of Testing and Materials, Philadelphia.

Morgan, E. L. *et al.* (1987) *Jour. of Environ. Sci.*, **30**, (2) pp. 47–49.

Morgan, E. L., Wright Jr, J. R. and Young, R. C. (1988) Developing a portable automated biomonitoring system for aquatic hazard evaluation. In J. Cairns (ed.) *Functional Testing for Hazard Evaluation.* STP series, American Society for Testing and Materials, Philadelphia.

Snedecor, G. W. and Cochran, W. G. (1980) *Statistical Methods,* 7th Ed., Iowa State University Press, Ames, Iowa, pp. 144–146.

SAS Institute (1979) *Statistical Analysis System (SAS) Users Guide.* SAS Institute, Cay, North Carolina.

Suarez, L. A., Barber, M. G., Lassiter, R. R. (1987) *GETS, a simulation model for dynamic bioaccumulation of nonpolar organics by gill exchange.* EPA/600/S3-86/057. US EPA, Environmental Research Laboratory, Athens, Georgia.

US EPA (1985) *Technical support document for water quality-based toxics control.* Office of Water, US Environmental Protection Agency, Washington, DC 74 pp. plus appendixes.

Part III

Biological systems using sensors other than fish

10

Enzyme microcalorimetry as a means of detecting organophosphates and carbamates in water

W. S. G. Morgan
National Institute for Water Research
P. C. Kühn
National Electrical Engineering Research Institute
Council for Scientific and Industrial Research, PO Box 395, Pretoria 0001, South Africa

10.1 INTRODUCTION

Industrial expansion, with the concomitant proliferation of chemicals, brings with it the danger that toxic substances will be released to surface waters in ever-increasing quantities. Such hazardous situations must be detected quickly to avoid deleterious effects upon the aquatic environment and public health. For many water pollutants there is no automatic monitoring equipment available. This has led aquatic toxicologists to investigate techniques whereby the overall impact of pollution upon living organisms may be assessed. In recent years equipment has been developed for the continuous, automatic monitoring of the response of sensor fish to a wide variety of toxic influences (Morgan, 1977, Cairns and Gruber, 1980; Cairns and van der Schalie, 1980). These systems have been successfully employed to monitor and control industrial effluent discharges to rivers and streams (Westlake and van der Schalie, 1976; Morgan *et al.*, 1982). However, living organisms have an inherent capability to adapt to very low levels of stress, and therefore do not, in all cases, respond to toxic concentrations deemed harmful to human health over the long term. A non-living toxic hazard-monitoring system could have many advantages in this respect.

One of the most important mechanisms of toxic action is thought to be the poisoning of enzyme systems. Research was therefore initiated to develop a continuous monitoring system based on the automatic detection of toxic substances which inhibit enzymes.

This chapter describes the use of a calorimetric technique to assess the inhibitory effects of organophosphorus and carbamate pesticides upon the enzyme cholinesterase, and discusses its application to environmental quality control.

10.2 THE DETECTION PRINCIPLE

Enzyme activity was determined by the use of flow microcalorimetry (Danielsson *et al.*, 1979) where the heat generated by a reacting substrate solution flowing through a calorimeter containing an immobilized enzyme preparation was measured. It has been shown experimentally (Monk and Wadsö, 1969) that the heat output rate for an enzyme reaction that is zero-order in substrate will give a recorded signal deflection proportional to enzyme activity. Enzyme inhibition may, therefore, be determined by a reduction in heat output rate.

The chemical reaction used to determine the activity of the immobilized enzyme was:

$$\text{butyrylcholine iodide} \rightarrow \text{butyric acid} + \text{choline iodide}$$

The enzyme (Horse-serum cholinesterase) hydrolyzed butyrylcholine three times as fast as acetylcholine, hence the use of the former as substrate (20 mM, prepared in phosphate buffer). Beezer and Stubbs (1973) stated that butyric acid reacts with a buffer solution in such a way that the enzyme/substrate reaction enthalpy is supplemented by the enthalpy of neutralization. Therefore, the heat output rate may have been enhanced through secondary reactions.

10.3 THE DETECTION DEVICE

The monitoring system comprised two main elements: an immobilized enzyme sensor unit and a data-acquisition incorporating control functions for the cyclic injection of test solutions (Fig. 10.1).

10.3.1 The enzyme sensor

The miocrocalorimeter employed (Fig. 10.2) was similar to that described by Danielsson *et al.* (1976), consisting of a glass column of 2 ml bed volume containing the immobilized enzyme preparation. The column was placed in a perspex barrel, leaving a small airspace between the column and the barrel for thermal insulation, ana arrangement which also simplified the interchange of columns. Cotton wool was used as a a bed support as well as being packed into the top of the column around the thermistor to hold the bed in position. A negative temperature coefficient thermister probe (Phillips No. 2322-627-1142, 4700 Ω) was mounted in a remobvable perspex holder which fitted into the barrel in such a way that the probe was situated well into the packed column bed, so as to give an even temperature signal. All joints were O-ring sealed. A heat exchanger, constructed from stainless steel capillary tubing (internal diameter 0.75 mm; length 250 mm) was placed immediately before the column in order to thermally equilibrate solutions before entry into the column. The unit stood on a stainless steel cylinder which enclosed the thermal equilibration coil,

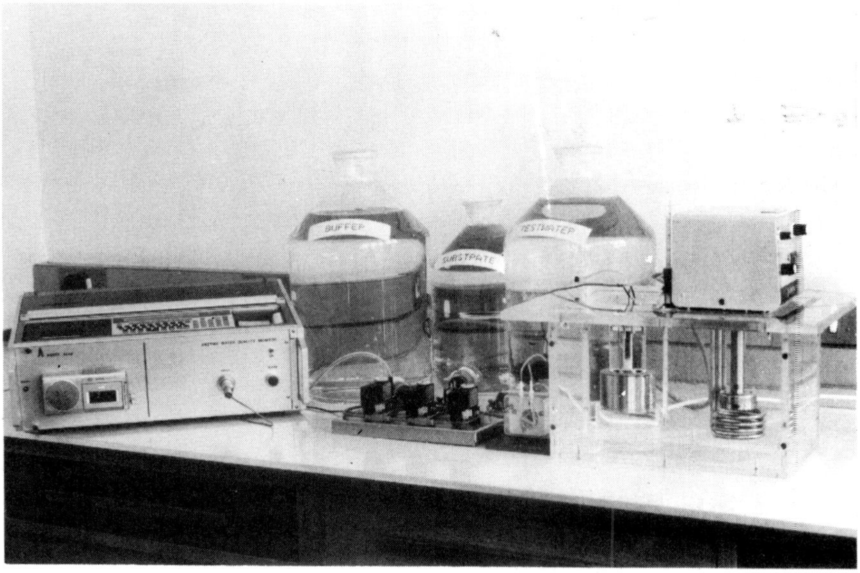

Fig. 10.1 — Equipment set-up showing (from left to right) electronic data acquisition and control module, test solutions (behind) and solenoid valve array (front), and enzyme micro-calorimeter in temperature-controlled water bath.

providing added thermal stabilization. The apparatus was almost totally immersed in a water bath maintained at a temperature of 27±0.02°C by an immersion circulator (Model P, Julabo Labortechnik, Seelbach, FDR). Buffer, substrate and test solutions were pumped through an automatically controlled solenoid valve array at a flow rate of approximately 2 ml/min. The temperature of a freshly prepared column stabilized within 30 min when buffer was allowed to flow continuously through the enzyme preparation.

The enzyme preparation was immobilized onto controlled pore glass (CPG), a support which is well characterized, has a high enzyme-binding capacity, is stabie under pressure and is not attacked by micro-organisms. Glutaraldehyde and γ-aminopropyltriethoxysilane were used to activate the CPG-10 beads (120/200 mesh, 631 Å diameter). The preparation of the alkylamine glass and its subsequent derivatization were carried out according to the procedure of Weetall and Filbert (1974), using 2.5% glutaraldehyde in phosphate buffer (0.1 M, pH 7). Approximately 5.0 mg of enzyme (horse serum cholinesterase, 218 U/mg) were added to 2 g of the activated glass in phosphate buffer and coupling was allowed to continue for 2.5 h at 4°C.

10.3.2 Data acquisition and control

The heat reaction in the microcalorimeter, when the substrate is passed through the enzyme preparation, causes temperature peaks in the range 0.01 to 1.0°C. The sensor is a thermistor with a nominal value of 4.7 kΩ at 25°C. This device exhibits a

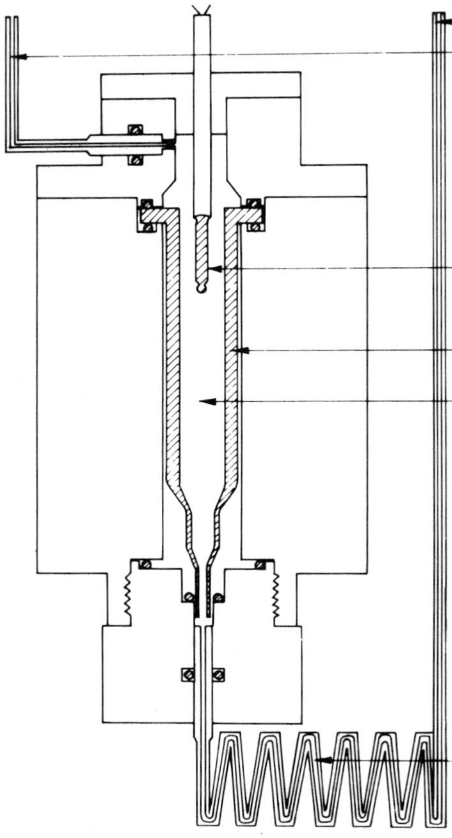

Fig. 10.2 — Enzyme microcalorimeter.

temperature coefficient of 125 Ω/°C for small excursions in temperature. This means that a temperature change of 0.1°C results in a resistance change of 12.5 Ω°C.

The resistance change is measured by passing a stable regulated current through the thermistor and measuring the voltage drop across it. This is carried out by a precision current source (Fig. 10.3). The change in voltage across the sensor is amplified by a differential chopper stabilized amplifier stage with a temperature stability of better than 40 parts in 10^6 per °C. This provides an output voltage change of 100 mV per °C. It was found that fluctuations due to noise amounted to the equivalent of 0.003°C. Because of the high sensitivities involved, the circuitry exhibits baseline drift owing to small fluctuations in the operating temperature. This is overcome by means of an automatic baseline correction scheme whereby a miocrocomputer controls a 12-bit digital-to-analog converter (DAC), whose output provides a compensating signal to control the zero offset of the amplifier. The computer, by sending instructions to the DAC, can thus achieve automatic baseline correction (Kühn and Morgan, 1983).

Accordingly, the data acquisition and control functions fall into five general categories:

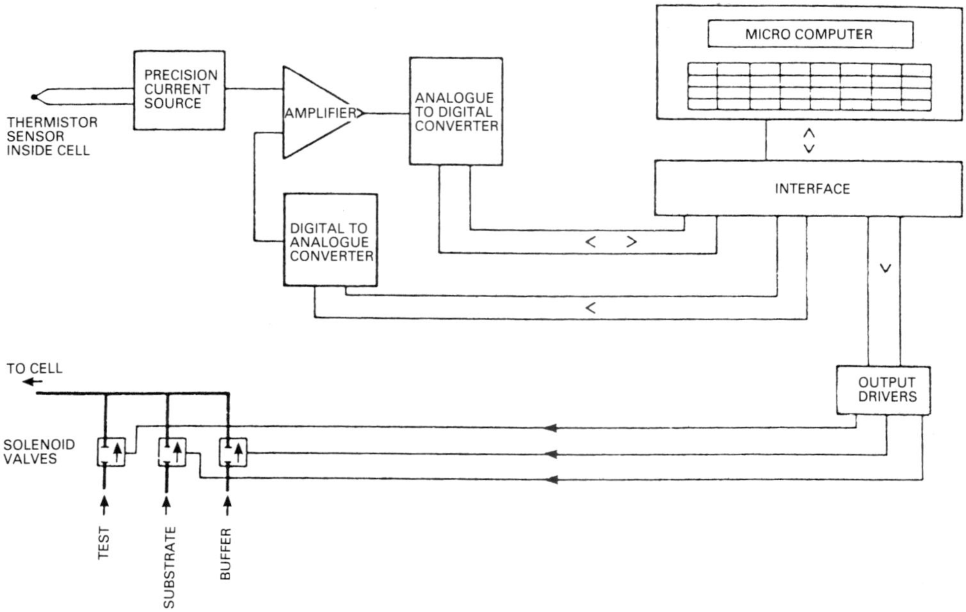

Fig. 10. 3 — Functional diagram.

- analog-to-digital conversion of the processed thermistor signal,
- digital-to-analog conversion of the auto-zero signal,
- timed control of the three solenoid valves,
- keyboard entry of certain test parameters,
- manipulation of measured data and printout of results.

These functions were all realized by using a programmable data acquisition and control until recently developed at our laboratories for general-purpose use in a research environment. The unit consists of an eight-bit analog-to-digital converter (ADC), a microcomputer with keyboard and liquid crystal display (programable in BASIC), a four-colour graphics printer/plotter and a cassette tape recorder for program and data storage.

The unit was customized by mounting the analog and DAC circuitry, together with their power supplies, inside the unit and configuring three 230 V output driver modules to actuate the three solenoid valves, and one 12 V d.c. module to activate an external alarm.

The output signal from the analog circuitry is applied to an eight-bit ADC. This gives a basic resolution of 0.4% of full scale. The input sensitivity of the ADC can be set as desired — it is at present configured for 250 mV full scale. The amplitude of the analog signal can thus be measured by the computer at any time via the interface and ADC, the smallest detectable temperature change being 0.01°C with the present settings.

A 12-bit DAC is also connected to the computer via the interface and its analog output is used to provide the automatic baseline correction (auto-zero).

Three a.c. solid-state relay modules provide for control of the three solenoid valves under computer control, and a fourth d.c. output module allows for activating the external alarm. A cassette tape recorder is incorporated in the instrument for program and data storage.

When the program is started, a menu appears on the display panel giving the user a choice of functions. These include 'NML', which executes a normal program; 'TIM', which allows for changing the solenoid valve timing parameters; 'CHT', which executes the normal program, and in addition plots a chart of the temperature peak while it is being monitored; and 'TAPE', which allows for data storage on cassette tape.

If the 'TIM' function is selected, the user is prompted through the process of determining the existing timing parameters and changing them, if required. Should either 'NML' or 'CHT' be selected, the user is prompted to enter certain parameters on the keyboard. These include data; type of toxicant and its concentration; and any general remarks about the specific test batch, as well as the required scaling factor.

The microcomputer initiates a test series as soon as the requisite parameters have been entered, and performs the automatic baseline correction routine while the test solution is passing through the cell. When the substrate solution passes through the cell, the computer begins to take a series of 100 samples at 2-s intervals of the peak amplitude in millivolts, which it stores in memory. If the 'CHT' function has been selected, then each value will also be plotted on a calibrated X–Y axis to form a chart recording of the temperature peak.

When the sampling period is over, the computer again measures the baseline value. If this has drifted from the initial residual value after auto-zero, the slope of the drift is automatically computed. The 100 sampled values are used to calculate the peak area which is corrected for residual baseline error and drift by the computer. The corrected peak area is then printed out, together with the number of the test. The area is printed in millivolt seconds.

At the end of each measurement cycle, the program returns to a point indicated on the flow diagram (Fig. 10.4) in order to continue with the next series of tests.

Throughout the above operations, messages are flashed onto the display panel to inform the user of the program status as well as the values of the measurements being taken.

10.4 DETECTION PROCEDURE AND SYSTEM EVALUATION

The test solution was allowed to flow through the enzyme column for a period of 5 min, followed immediately by the substrate, a 30 s pulse of butyrylcholine iodide (20 mM). Phosphate buffer was then used to flush the column until the temperature response, due to the heat of reaction, as measured by the signal output, had dissipated. This was indicated by the re-establishment of a stable baseline, a procedure which was accomplished in 3.5 min, providing a full cycle of 9 min. The cycle was repeated continuously, peak heights and areas being measured and graphically plotted against exposure time. An alarm status was initiated at a 50% decrease in original heat response. The inhibitory effects of Metasystox® [O-O-dimethyl *S*-2-(ethylsulphinyl) ethyl phosphorothioate]; DDVP [2,2-dichlorovinyl dimethyl phosphate]; Lebaycid [50% Fenthion-O,O dimethyl O-[4-methylthio]-*m*-

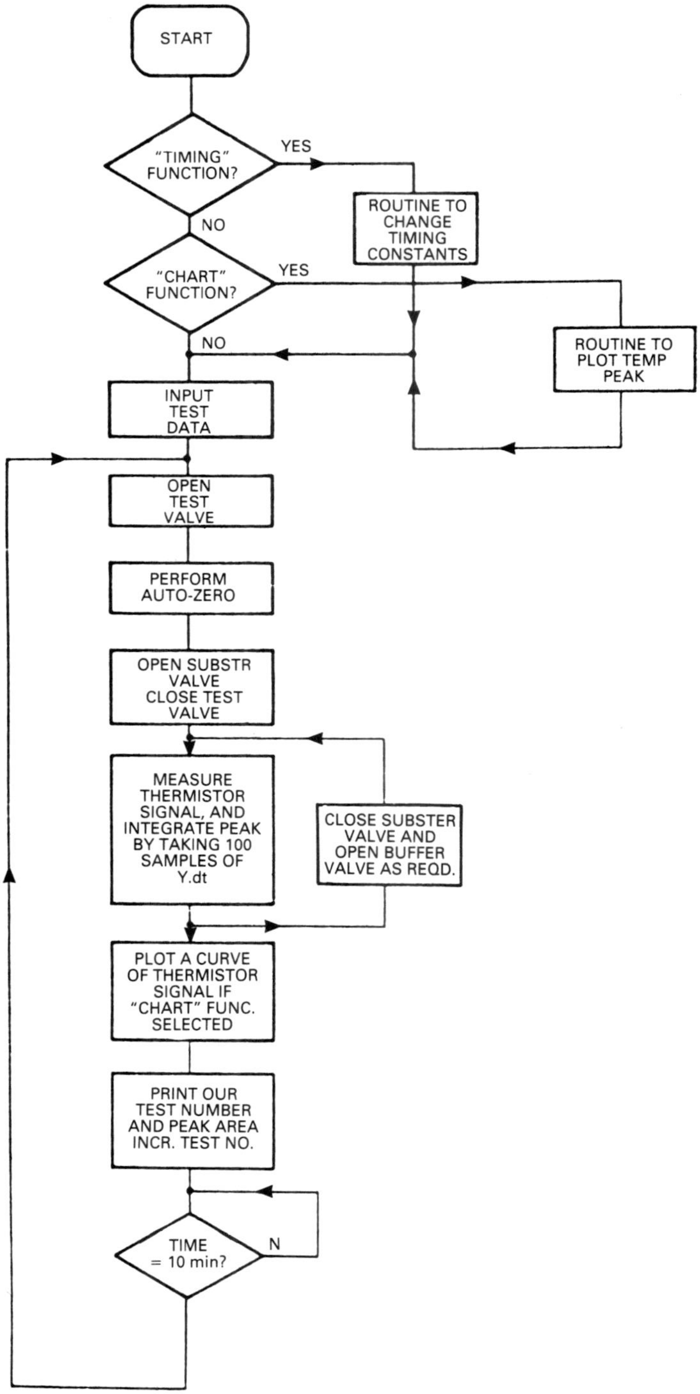

Fig. 10.4 — Flow diagram.

tolyl) phosphorothioate]; Carbaryl [1-naphthyl methyl carbamate] and Carbofuran [2,3-dihydro-2,2-dimethyl-benzofuran-7-yl methyl cartbamate], on cholinesterase, were studied. The pesticides were used as pure formulations (Polyscience, Niles, Illinois) dissolved in hexane, a solvent which had no inhibitory effects upon cholinesterase under control conditions.

Subsequent to thermal equilibration of the enzyme preparation, an assay scheme such as that depicted in Fig. 10.5 was used to assess the inhibitory effects of the various nominal pesticide concentrations employed. A series of approximately 10 cycles was carried out to establish a 100% activity value for the enzyme preparation, exposed to solvent only. Thereafter a nominal pesticide concentration was established in the test solution and cyclic injection of test, substrate and buffer continued until enzyme activity had decreased to 50% of its original value.

Inhibition of the cholinesterase was reflected by the lower response to the substrate pulse following administration of the test sample. Enzymatic degradation could be assessed by a visual observation of peak height. However, it was found that integration of the temperature peaks provided a better linear relation to substrate concentration and supplied more accurate data than peak height.

Fig. 10.6 depicts the inhibitory effects of various pesticide concentrations on cholinesterase activity. All pesticides inhibited enzyme activity to a greater or lesser extent with respect to exposure time, depending upon the concentration employed. The fact that higher concentrations inhibited enzyme activity in a shorter period of time would be of advantage when continuously assessing water quality. It would be possible, in an operative monitoring situation, to program the microcomputer to assess the slope of an inhibition curve to provide an estimate of hazard.

The large number of cycles required for detection, at 50% enzyme inhibition, of very low concentrations of pesticides (Table 10.1) may be considered as providing too great a time-lag for effective monitoring. Although the alarm status may be set at any arbitrary degree of enzyme inhibition, the degree of sensitivity established by producing an automatic alarm at the 50% enzyme inhibition level would, in our opinion, be adequate for the protection of individuals using water supplies. For example, the LD_{50} of DDVP for man has been assumed to be 6 mg/kg (Heath, 1961). A man weighing 70 kg would need to consume 4200 l of water containing 0.1 mg/l of DDVP to receive the LD_{50} dose in the time taken to detect that quantity (approximately 24 h).

10.5 CONCLUSIONS

The preliminary investigation has demonstrated the feasibility of developing a monitoring system which will automatically detect many hazardous substances present in the environment. Initial experimentation has suggested that specific enzymes may be utilized to detect heavy metals (urease, invertase), phenolic compounds (tyrosinase), cyanide (rhodenase) and carcinogenic aromatic amines (horseradish peroxidase). A search for enzymes inhibited by specific toxicants is a logical step in the future development of the monitoring system. The enzymes selected could be immobilized and incorporated into a multi-enzyme detection system which may afford greater analytical specificity than can be achieved using biological sensors. In this context it should be pointed out that after the detection of a

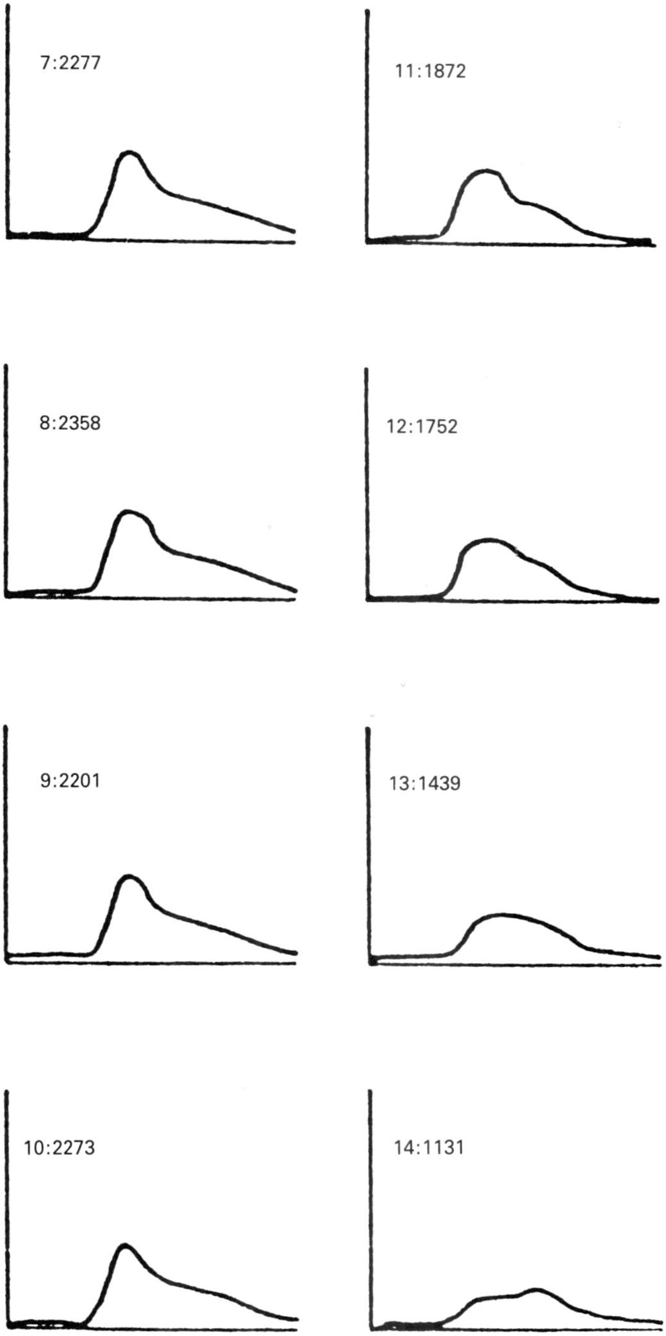

Fig. 10.5 — Calorimetric records before (7–10) and after (11–14) exposure of cholinesterase to 10 mg DDVP/l.

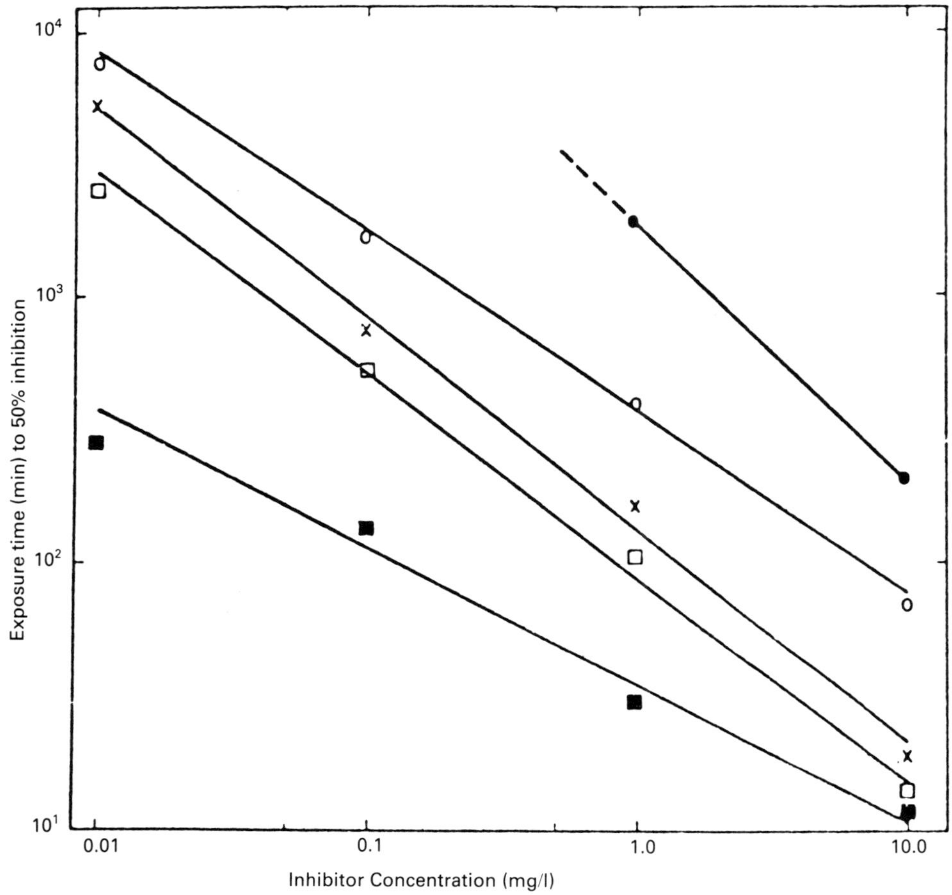

Fig. 10.6 — Inhibition of cholinesterase by Metasystox R (○), DDVP (+), Lebaycid (○–○), Carbaryl (●) and Carbofuran (■). Log exposure time to a 50% decrease in enzyme activity is plotted versus log inhibitor concentration.

changed temperature signal the nature of the inhibiting material could be determined by conventional analytical procedures, such as atomic absorption spectroscopy, as it will remain bound to the enzyme column material.

Utilization of a microcomputer has made the instrumentation sufficiently flexible to cope with the changing demands of such a research program. It is quite possible to totally reconfigure the machine by changing the software. This can be done directly from the keyboard on the instrument by anyone with a knowledge of BASIC and the parameters of the system. It is also possible to add software to perform data analyses, for example to compare multiple sets of data, which should prove most useful once a database has been compiled.

ACKNOWLEDGEMENTS

We wish to acknowledge the contributions made by A. G. E. Webb and C. G. Hulme (National Electrical Engineering Research Institute (NEERI)) in the development

Table 10. 1 — Relationship between exposure time to 50% enzyme inhibition (ET_{50}) and pesticide concentration

Pesticide	Chemical type	Conc. (mg/l)	ET_{50}	No. of cycles
Metasystox R	Organophosphate	10.1	13.8	3
		1.0	104.7	21
		0.1	512.8	103
		0.01	2511.9	502
DDVP	Organophosphate	10.0	18.7	4
		1.0	161.0	33
		0.1	714.0	143
		0.01	5248.0	1050
Lebaycid	Organophosphate	10.0	71.0	15
		1.0	385.0	77
		0.1	1666.0	333
		0.01	7716.0	1543
Carbaryl	Carbamate	10.0	200.0	40
		1.0	1842.0	369
Carbofuran	Carbamate	10.0	12.2	2
		1.0	30.5	6
		0.1	132.0	27
		0.01	277.0	55

of the data acquisition and control module, A. J. Hassett (National Institute for Water Research (NIWR)(for valuable advice on pesticide formulations and C. J. van Niekerk (NIWR) for assistance with the analyses. The paper is published with the approval of the Chief Directors of the NEERI and NIWR, whom we thank for their encouragement and support.

REFERENCES

Beezer, A. E. and Stubbs, C. E. (1973) Application of flow microcalorimetry to analytical problems. 1. Determinaation of organophosphorus pesticides by inhibition of cholinesterase. *Talanta* **20** 27–31.

Cairns Jr, J. and Gruber, D. (1980) A comparison of methods and instrumentation of biological early warning systems. *Wat. Resour. Bull.* **16** 261–266.

Cairns Jr, J. and van der Scalie, W. H. (1980) Biological monitoring. Part 1. Early warning systems. *Wat. Res.* **14** 1179–1196.

Danielsson, B., Gadd, K., Mattiasson, B. and Mosbach, K. (1976) Determination of serum urea with an enzyme thermistor using immobilized urease. *Analytical Letters* **9** 987–1001.

Danielsson, B., Mattiasson, B. and Mosbach, K. (1979) Enzyme thermistor analysis in clinical chemistry and biotechnology. *Pure and Appl. Chem.* **51** 1443–1457.

Heath, D. F. (1961) *Organophosphorus poisons*. Pergamon Press, New York.

Kühn, P. C. and Morgan, W. S. G. (1983) Monitoring water quality by the electronic detection of enzyme inhibition. In D. Gilbert and M. Crabtree (eds) *Power and control in water management*, Vector/Pulse, South Africa.

Monk, P. and Wadsö, I. (1969) Flow microcalorimetry as an analytical tool in biochemistry and related areas. *Acta. Chem. Scand.* **23** 29–36.

Morgan, W. S. G. (1977) An electronic system to monitor the effects of changes in water quality on fish opercular rhythms. In: J. Cairns Jr, K. L. Dickson and G. F. Westlake (eds) *Biological Monitoring of Water and Effluent Quality*, ASTM STP 607, pp. 38.55, American Society for Testing and Materials, Philadelphia.

Morgan, W. S. G., Kühn, P. C., Allais, B. and Wallis, G. (1982) An appraisal of the performance of a continuous fish biomonitoring system at an industrial site. *Wat. Sci. & Tech.* **114** 151–161.

Weetall, H. H. and Filbert, A,. M. (1974) Porous glass for affinity chromatography applications. In: *Methods in Enzymology* **34** pp. 59–72.

Westlake, G. F. and van der Schalie, W. H. (1976) Evaluation of an automated biological monitoring system at an industrial site. In: J. Cairns Jr., K. L. Dickson and G. F. Westlake (eds) *Biological Monitoring of Water and Effluent Quality*, ASTM STP 607, pp. 30–37, American Society for Testing and Materials, Philadelphia.

11

A real-time computerized video tracking system to monitor locomotor behavior

Robert M. Korver† and **John B. Sprague**,
Department of Zoology, University of Guelph, Guelph, Ontario, Canada, N1G 2W1

11.1 INTRODUCTION

Recent developments in computerized video-processing technology have greatly enhanced our ability to make automated measurements of locomotion in animals. Automation eliminates manual observation of animals and tedious data transcription. This chapter describes a new system originally used in experiments to make classical laboratory assessments of pollutant avoidance in fish more predictive to the field (Korver, 1986; Korver and Sprague, in press). The system features: (a) inexpensive off-the-shelf technology consisting of a standard, black-and-white video camera and monitor attached to a digitizing board installed in a personal computer; (b) real-time processing — results are immediately available for viewing and analysis; (c) adaptability to a variety of experimental situations, e.g. tracking movements of a small organism on a microscope slide or a large terrestrial organism; (d) the ability to allow the subject to 'disappear' behind an occlusion; and (e) menu-driven operation.

11.2 HARDWARE REQUIREMENTS

The tracking system (Fig. 11.1) can use almost any good quality black-and-white camera with standard video output. The system described here used a Panasonic WV-1550 TV camera with a Tokina 12.5–75 mm f1.8 6× zoom lense. This camera is sensitive to low light levels and minimizes 'burn-in' and 'ghosting'. Almost any type of video monitor may be used, and the inexpensive Zenith ZVM-1230-A composite computer monitor was found to be adequate.

† Present address: Fisheries Branch, Ontario Ministry of Natural Resources, PO Box 50, Maple, Ontario, Canada, L0J 1E0.

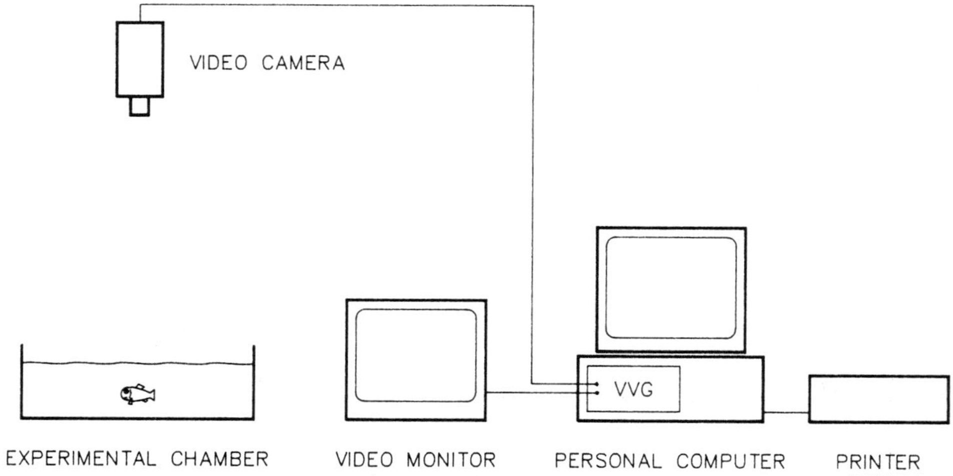

Fig. 11.1 — Basic components of the tracking system. The video monitor allows visual observation of the subject in the experimental chamber.

The minimum computer configuration is an IBM Personal Computer (PC) with 256 K of memory, two 360-K disk drives, and color/graphics adapter. A color monitor is most essential as the programs use color graphics quite extensively. Almost any compatible dot-matrix printer with 132 characters/line and graphics capability will suffice.

The key component of the tracking system is the PC-Mate™ Video Van Gogh (VVG) digitizing board, made by Tecmar, which fits in an expansion slot on the PC. This board digitizes all or part of a video image up to a maximum resolution of 256×256 pixels. Each pixel is addressable and may be assigned a grey value ranging from 0 to 255, corresponding to the brightness of the video image at that location.

11.3 SOFTWARE DESCRIPTION

The tracking software works on the principle of distinguishing an object from a contrasting background, and assigning (x, y) coordinates to its location. Various measures of locomotor behavior are calculated from sequential locations and these can be displayed, printed, and converted to files readable by most computers and statistical packages.

The software is written as a series of IBM BASIC programs running under PC-DOS Version 2.0 or higher. Macro Assembler routines are called to digitize and display images, and to track the subject. The programs comprise some 2500 lines of BASIC and 2000 lines of Assembler. Some routines from the VVG Software Support Package, Version 1.1, are incorporated into the tracking software.

There are four main menus for tracking, data display, data summary, and statistical formatting, described below.

11.3.1 Tracking menu

The first nine items of the tracking menu (Fig. 11.2) allow the user to calibrate the

```
                    TRACKING MENU

                Press the desired F-key.

F1.   Take full picture
F2.   Display full picture and set colour mapping
F3.   Enter scanning and shelter boundaries (do F1 first)
F4.   Calibrate grey levels of subject (do F1 first)
F5.   Set density, window size and options for scanning
F6.   Set time between scans
F7.   Set number of scans between data records
F8.   Write calibration parameters to file
F9.   Read calibration parameters from file
F10.  Track subject
Ctrl F2. Data display menu
Ctrl F3. Data summary menu
Ctrl F4. Statistical formatting menu
Ctrl F5. End program
```

Fig. 11.2 — Screen display of tracking menu.

system for actual tracking, which is initiated through item 10.

Menu items 1 and 2 allow digitization and viewing of a calibration image. The image can be displayed in four colors corresponding to ranges of grey values; colors and ranges may be changed to enhance or mask certain features of the image.

Menu item 3 establishes the boundaries of the rectangular scanning or observation area. Boundaries for a shelter or occlusion behind which the subject can disappear may also be entered. While the subject is under the shelter, the (x, y) coordinates for the center of the shelter are taken as the location of the subject. The location of a warning lamp outside of the scanning area may also be entered. The warning lamp can be used as an event marker, say for designating the presence or absence of a stimulus, or for procedural problems such as low water-levels.

Menu item 4 allows the user to select a range of grey values for the subject. Portions of the image outside this range will subsequently be taken as background during tracking (compare parts (a) and (b) of Fig. 11.3).

The main purpose of menu item 5 is to optimize the speed of scanning during tracking by specifying the types of scans to be performed. A drawback of the VVG board is the relative slowness with which it digitizes an image. Images are scanned from left to right, manifested on the video monitor by a white, vertical line. A full resolution image (about 256×256 pixels) takes approximately six seconds to digitize, not counting any processing dedicated to locating the subject. This creates problems with a fast-moving subject since its image may be compressed or lengthened depending on direction of travel relative to the scan. The density of resolution of the scan can range from digitizing every pixel (much too slow for most applications) to

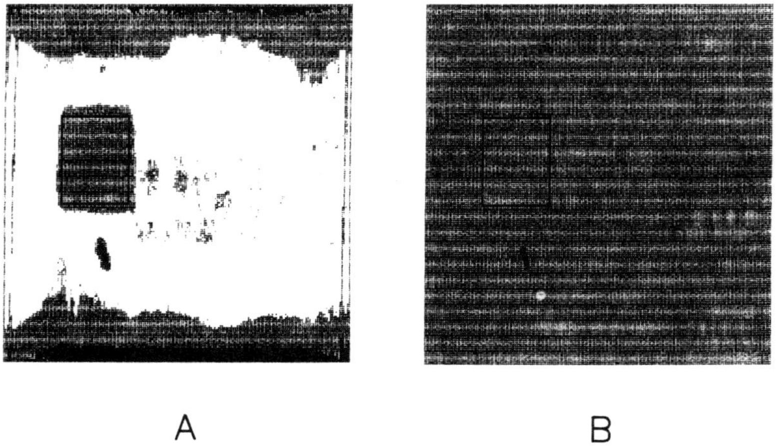

Fig. 11.3 — Sample screen display during selection of a range of grey values that distinguish the subject (a fish) from its background. A shelter has been defined in the left side of the observation area. A. Raw image displayed in four colors (shades of grey here). B. Final image displayed here has no pixels on the background within the range selected for the subject.

digitizing every sixth pixel horizontally and vertically. The density chosen will depend on the size of the subject, its maximum velocity, and the resolution of results desired. Scanning every second pixel horizontally and vertically gave satisfactory speed and results for the avoidance experiments.

While tracking, the program searches for all pixels having grey values within the range established for the subject. The median x and y coordinates for these pixels are then taken as the location of the subject. However, the whole scanning area need not be digitized to locate the subject. A scanning window that centers on the location of the subject in the previous scan allows a much higher scanning frequency. Window sizes may range from a few pixels on a side to the full scanning area. The size is chosen so that the subject will not normally be able to move outside the window between successive scans.

From the above description, it should be obvious that the program is not intended to track several subjects independently. It can, however, find an overall median 'location' for a group of distinct subjects.

The program 'loses track' of the subject when less than two pixels in the scanning window are within the range of grey values selected in menu item 4. Several options are available for this condition. If no shelter is present, the user may elect to have tracking terminated or to have the program continue by performing scans of the whole observation area until the subject is found. Normal tracking is then resumed. If a shelter is present, the program may be set to scan an area around the shelter, or to scan the entire observation area rather than a window. A location of (0,0) is assigned when the program cannot locate the subject and this is treated as a missing value. The coordinates for the center of the shelter are taken as the location of the subject under these circumstances. When the subject is found, normal tracking is resumed.

Menu item 6 is used to set the time interval between scans. With a small window size, the program can scan at intervals down to about 0.2 s. The maximum allowable interval between scans is 59.99 s.

For each scan, the program finds the location of the subject and the pythagorean distance (in pixel units) that it travelled from its last recorded location. Although the time interval between scans may be short to keep track of the subject, it is usually not necessary to generate a data record on disk for every scan. Menu item 7 allows the user to establish how often data records are written to disk (up to a maximum of 65536 scans/data record).

The calibration parameters described above may be stored on disk, and retrieved, through menu items 8 and 9, respectively. Thus, repeated trials for a particular application are more quickly initiated, usually requiring little or no fine tuning of the stored parameters.

Once calibration is complete, tracking is initiated through menu item 10. Information to document the session is entered, as is a file-name for the data, the scheduled duration of tracking (up to 100 hours or 32767 data records) may be subdivided into control, experimental, and recovery periods. If an experimental period is scheduled, warning beeps will be activated two minutes before it is to begin and end. A timer feature allows tracking to begin automatically up to 24 hours later.

During tracking, a status screen as in Fig. 11.4 appears and is updated for every

```
                         TRACKING STATUS

Toxicant:          Zinc
Concentration:     2000 ppb

                         Scheduled          Actual

Tracking start time:         -              01:32:03:43
Toxicant start time:     04:32:03:43
Toxicant stop time:      07:32:03:43
Tracking stop time:      10:32:03:43

Present time:  01:34:31:43   Elapsed time:  00:02:28:00
Present location:   ( 57,172)




1. Toxicant started   2. Toxicant stopped   3. Write to disk   4. Stop tracking
```

Fig. 11.4 — Sample screen display during tracking. This screen was copied 2 minutes and 28 seconds into a nine-hour session.

scan. Scheduled and actual times for the various periods comprising the session are displayed, as are current and elapsed time. Times are displayed as hours:minutes-:seconds:hundredths of seconds. The present location of the subject in pixel coordinates is also shown. Messages requiring possible user intervention (e.g. two-minute warning for experimental period, warning lamp on) are displayed below the location of the subject (none are shown in Fig. 11.4). Tracking is automatically terminated at the scheduled time or may be stopped at any time by the user.

11.3.2 Data display menu

Preliminary examination of data is done through the data display menu (Fig. 11.5).

```
                DATA DISPLAY MENU

              Press the desired F-key.

F1.   Enter filename (DO FIRST)
F2.   Display documentation file on screen
F3.   Print documentation file
F4.   Display message file on screen
F5.   Print message file
F6.   Display data file on screen
F7.   Print data file
F8.   Plot movement of subject
Ctrl F1.   Tracking menu
Ctrl F3.   Data summary menu
Ctrl F4.   Statistical formatting menu
Ctrl F5.   End program
```

Fig. 11.5 — Screen display of data display menu.

Three files are generated in a given tracking session: a documentation file, a message file; and a data file. Each of these may be sent to the screen or the printer through menu items 2 through 7 (Fig. 11.6 gives a typical printer output).

The documentation file contains the settings of all calibration parameters in effect for a given tracking session. The message file shows the time of occurence of various events in a tracking session. Message-generating conditions requiring immediate user-intervention would have been displayed on the tracking status screen (Fig. 11.4) with warning beeps if necessary.

Each record in the data file gives the elapsed time since the beginning of the session, the location of the subject at that instant, and the distance travelled in pixel units since the previous record, Distance travelled is cumulated for every scan, rather than calculated once for every data record. This gives greater accuracy when many scans are done between data records.

Menu item 8 allows preliminary examination of the data by plotting locations visited in the observation area by the subject for all or part of a session (Fig. 11.7). Points may be joined by lines to trace the subject's path and may be color-coded to indicate observations taken during control, experimental, and recovery periods, or observations taken when the warning lamp was lit. This feature is useful to gain a quick, subjective impression of the subject's preferred areas in the observation chamber under various experimental conditions. Caution must be exercised, however, as specific locations visited more than once appear as only one point.

DOCUMENTATION FOR: 12035003

Date: 12-12-1985
Toxicant: Zinc
Concentration: 3500 ppb

Scanning boundaries (upper left and lower right corners): (48, 63) (228,244)
Warning lamp coordinates: (20, 84)

Scanning Option: 2
'Tracking interrupted for full scan' message suppressed.

Lower threshold: 0 Upper threshold: 85
Window width: 44 Window height: 56
Scanning density: 4
Scanning time: 50 centiseconds
Shelter or Option 2 scanning time: 250 centiseconds
Scans between data summaries: 30

Duration of tracking (baseline): 180 minutes
Duration of tracking (with toxicant): 180 minutes
Duration of tracking (after toxicant stopped): 180 minutes

MESSAGES FOR: 12035003

Actual Time	Elapsed Time	Message
11: 0: 1:72	0: 0: 0: 0	Tracking started
14: 0: 1:72	3: 0: 0: 0	Toxicant started
17: 0: 1:72	6: 0: 0: 0	Toxicant stopped
20: 0: 2:22	9: 0: 0:50	Tracking stopped

DATA FOR: 12035003

Record	Time	X	Y	Distance
1	0: 0:15: 0	50	233	423
2	0: 0:30: 0	52	87	435
3	0: 0:45: 0	48	191	602
4	0: 1: 0: 0	54	211	515
5	0: 1:15: 0	66	109	389
6	0: 1:30: 0	56	79	473
7	0: 1:45: 0	52	207	601
8	0: 2: 0: 0	50	203	526
9	0: 2:15: 0	62	65	396
10	0: 2:30: 0	68	71	44

Fig. 11.6 — Sample printer output of the documentation, message, and data files. No shelter was present for this tracking session, which was nine hours long. Each line of the output for the data file shows a data record generated at 15-second intervals. Only the first 10 of 2160 data records are shown.

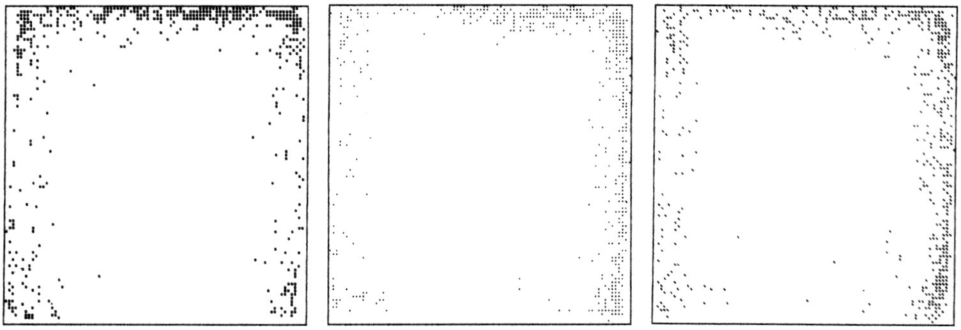

A B C

Fig. 11.7 — Pictorial representation of locations visited by a fish at 15-second intervals. Data from a control period (panel A), an experimental period (Panel B), and a recovery period (panel C) are displayed separately for clarity. In B, a toxicant flowed through the left half of the observation area, and there was an obvious avoidance response.

11.3.3 Data summary menu

The data summary menu (Fig. 11.8) converts the raw data records collected during

```
                DATA SUMMARY MENU

             Press the desired F-key.

F1.   Enter filename (DO FIRST IF USING F2-F8)
F2.   Define areas
F3.   Define intervals between data summaries
F4.   Define units and scale
F5.   Set criteria for turns and resting
F6.   Write summary parameters to file
F7.   Read summary parameters from file
F8.   Generate data summary file
F9.   Display data summary on screen
F10.  Print data summary
Ctrl F1. Tracking menu
Ctrl F2. Data display menu
Ctrl F4. Statistical formatting menu
Ctrl F5. End program
```

Fig. 11.8 — Screen display of data summary menu.

tracking to a variety of quantitative measures of locomotor behavior. These statistics are stored in a disk file from where they may be sent to the screen or printer or reformatted into ASCII files through the statistical formatting menu.

For a given data set, up to 16 rectangular areas may be defined within the observation area through menu item 2. These areas may be labeled in terminology pertinent to the experiment. For example, in avoidance experiments, the observation area may be divided into two equal areas corresponding to the 'dosed' and 'clean' sides of the test tank. For other applications, areas may overlap, or be combined for odd shapes. Undefined portions of the observation area are assigned to an additional 'remainder' area, so the test chamber could, say, be partitioned into a rectangular central area and a remainder area on the periphery.

Raw data are summarized over user-definable time periods (menu item 3). Data summary intervals may be equally or unequally spaced in time, but must be at least one minute in duration.

The program will calculate, for a given summary interval, the number and percentage of observations of the subject in each of the 17 possible areas. An observation corresponds to one record in the data file generated during tracking. The number of trips to each area and the mean duration of each trip are also calculated. If a shelter is present, these statistics are also calculated for observations in and out of the shelter.

To convert measures of distance travelled from pixels to more conventional units, the user must enter the desired units (e.g. mm, cm, m) and the width of the observation area in those units (menu item 4). Within the constraints of the distance threshold described below, this allows calculation of total distance travelled and mean velocity for a given summary interval.

Two threshold values are required before calculations are done. The first defines the minimum angle in pi radians that the subject must deviate from its previous trajectory in order for a 'turn' to be counted. A threshold angle must be set because the high resolution of the tracking system will almost always cause a change in bearing to register, even if the subject is following an essentially straight path.

The second threshold value is the minimum distance the subject is required to travel between observations in order to distinguish between 'moving' and 'resting'. Even for stationary subjects, the tracking system often returns slightly different locations for successive scans, thereby adding some background noise to the data. Setting a distance threshold eliminates most of this noise, and allows the user to mask out slight movements of the subject that are not significant to its overall locomotor behavior. Turns are only counted for 'moving' subjects.

Using the two threshold values, various additional statistics are calculated. These include: number of turns; accumulated and mean size of turns; total distance travelled; total and mean straight path distance travelled; and number and percentage of observations moving and resting. Turn statistics are found for left, right and total turns.

The selection screen for output of statistics (menu item 9 and 10) summarizes the calculations done by the program (Fig. 11.9). Items desired in the output are entered into the list at the bottom of the screen by moving a cursor to the desired variable (numbered from 1 to 100) and pressing carriage return. If a shelter is defined for the data summary file in use, selections for 'In shelter' and 'Out of shelter' are added under the headings for number and percentage of observations, number of trips, and mean time. Up to 10 variables at a time may be sent to the printer in 80 characters/line mode, or to the screen. Up to 18 may be sent if the printer is in 132 characters/line mode. Variables may be displayed across the screen or page in any order. A sample

```
            OUTPUT VARIABLE SELECTION FOR B:120350d3.sum (PRINTER)

                                    Number of Observations
[1] All areas [2] Missing [3] Moving [4] Resting
Area   1   2
     [7] [8]
                                    Percent Observations
[24] All areas [25] Missing [26] Moving [27] Resting
Area   1   2
     [30] [31]
                                    Number of Trips
[47] All areas
Area   1   2
     [50] [51]
                                    Mean Time
[67] All areas
Area   1   2
     [70] [71]
Number of Turns  Size of Turns (Pi radians)       Distance          Mean Velocity
[87] Left        [90] Acc. left   [93] Mean left    [96] Total        [99] cm/second
[88] Right       [91] Acc. right  [94] Mean right   [97] Total str.   [100] cm/minute
[89] L and R     [92] Acc. L and R[95] Mean L and R[98] Mean str.
Order:  1   2   3   4   5   6   7   8   9   10  11  12  13  14  15  16  17  18
Var.:  [7] [8] [2] [3] [4] [30][31][25][26][27][50][51][70][72][96][99]
→,←,↓,↑-Move   Ctrl→,Ctrl←-Order   C.R.-Select   B-Blank   P-Print   M-Menu
```

Fig. 11.9 — Screen display for selecting output statistics. Actual display is color-coded for greater clarity.

printer output for the variables selected in Fig. 11.9 (list of boxed numbers at the bottom of the screen) is given in Fig. 11.10. Printing is optional for the documentation and legend portions.

Although the program calculates all of the above-described statistics and writes them to disk (menu item 8), it is the user's responsibility to judge the usefulness of each statistic. The depiction of the data summary process in Fig. 11.11 shows in a somewhat exaggerated manner how, for example, details of turning behavior may be lost as the time interval between data records is increased. Only one turn would be counted for the five records shown, whereas if a data record had been generated for every scan, considerably more turns would be logged. Further comments on limitations of the tracking system, effects of sampling frequency and data artifacts are found in the section on sources of error below.

11.3.4 Statistical formatting menu

The statistical formatting menu (Fig. 11.12) allows statistics generated through the data summary menu to be selected for output into ASCII-format files. Data in these files may then be easily read by most statistical packages and/or transferred to other computers for further analysis.

Up to 80 data summary files may be merged into one ASCII file. A list of data summary filenames is entered in menu item 1, and the data records in the output file contain a field for the position of the input filename in the list. The individual tracking sessions comprising a given experiment may then be easily sorted into replicates, treatments, etc.

The user can select statistics of interest (menu item 2) from a variable selection

DOCUMENTATION FOR: B:120350d3.sum

Width of scanning area:	50 cm
Threshold angle:	0.08333333 pi radians
Threshold distance:	2 cm

AREA #	BOUNDARIES	LABEL	COMBINE WITH AREA #
1	(135, 63)-(228,244)	Clean	
2	(48, 63)-(134,244)	Dirty	

LEGEND FOR: B:120350d3.sum

COLUMN HEADING	INTERPRETATION
TIME	Elapsed time in minutes
NOA1	Number of observations in Area 1
NOA2	Number of observations in Area 2
NOMISS	Number of observations missing
NOMOVE	Number of observations moving
NOREST	Number of observations resting
POA1	Percent observations in Area 1
POA2	Percent observations in Area 2
POMISS	Percent observations missing
POMOVE	Percent observations moving
POREST	Percent observations resting
NTA1	Number of trips to Area 1
NTA2	Number of trips to Area 2
MTA1	Mean time in Area 1 (sec)
MTA2	Mean time in Area 2 (sec)
DT	Total distance travelled (cm)
MVS	Mean velocity (cm/second)

SUMMARY DATA FOR: B:120350d3.sum

TIME	NOA1	NOA2	NOMISS	NOMOVE	NOREST	POA1	POA2	POMISS	POMOVE	POREST	NTA1	NTA2	MTA1	MTA2	DT	MVS
180	407	295	18	708	12	56.5	41.0	2.5	98.3	1.7	34	35	180	126	35730	3.308
360	575	121	24	714	6	79.9	16.8	3.3	99.2	0.8	17	17	507	107	42731	3.957
540	506	164	50	717	3	70.3	22.8	6.9	99.6	0.4	24	23	316	107	45630	4.225

End of file.

Fig. 11.10 — Sample printer output of a data summary. The documentation gives the user-defined parameters used to make the calculations. The legend explains the headings used in the actual summary for the selected statistics.

screen similar to that of Fig. 11.9. A record layout (Fig. 11.13) may be displayed or printed through menu item 3, showing the starting position of each variable in the record. This information must be supplied to the program(s) reading the ASCII file.

11.4 SOURCES OF ERROR

Although the tracking system is capable of collecting detailed and accurate information on locomotor behavior, there are some sources of error. Three main

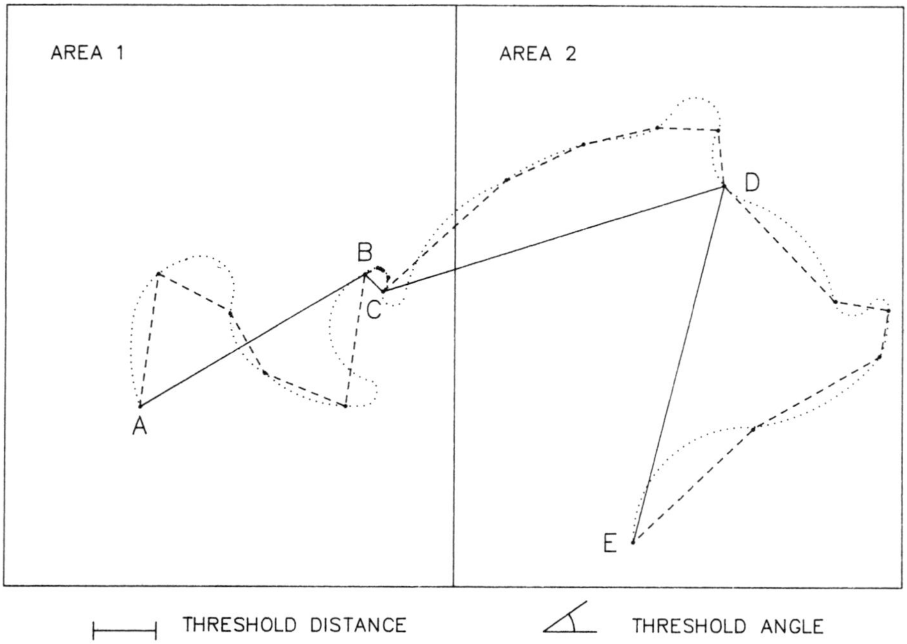

Fig. 11.11 — Pictorial representation of the data summary process, demonstrating potential loss of detail with relatively long time-intervals between data records. The dotted line represents a hypothetical true path for a subject. The dashed line indicates the pythagorean distance travelled between scans which is cumulated for every data record (lettered points along solid line). The program would count three observations in Area 1 and two observations in Area 2, with one trip to each area. The subject would be 'resting' between B and C as the distance along the dashed line is less than the threshold distance. This resting period causes the observations between A and D to be counted as a straight path, as CD deviates from the path defined by AB by less than the threshold angle. One right turn would be counted through observation D.

```
          STATISTICAL FORMATTING MENU

            Press the desired F-key.

     F1.  Enter filename(s) to merge
     F2.  Select variables
     F3.  Display/print variables selected
     F4.  Generate ASCII file
     F5.  Display files on drive B
     Ctrl F1. Tracking menu
     Ctrl F2. Data display menu
     Ctrl F3. Data summary menu
     Ctrl F5. End program
```

Fig. 11.12 — Screen display of statistical formatting menu.

VARIABLES SELECTED AND RECORD LAYOUT FOR ASCII FILE

Field No.	No. of Bytes	Starting Position	Description
0	6	1	Elapsed time in minutes
1	6	7	Number of observations in Area 1
2	6	13	Number of observations in Area 2
3	6	19	Number of observations missing
4	6	25	Number of observations moving
5	6	31	Number of observations resting
6	9	37	Percent observations in Area 1
7	9	46	Percent observations in Area 2
8	9	55	Percent observations missing
9	9	64	Percent observations moving
10	9	73	Percent observations resting
11	6	82	Number of trips to Area 1
12	6	88	Number of trips to Area 2
13	9	94	Mean time in Area 1 (sec)
14	9	103	Mean time in Area 2 (sec)
15	9	112	Total distance travelled
16	9	121	Mean velocity (/second)
17	3	130	Position in list of filenames

Number of fields: 18 Record length: 132

Fig. 11.13 — Sample printer output showing the layout of each record in the data summary file after conversion into ASCII format. This information is required for other programs and/or computers to read the file.

categories are errors attributable to hardware, those arising from the algorithms used for tracking, and those associated with calculation of statistics. Those categories are not necessarily mutually exclusive.

The main source of variation in the hardware is differences in grey-level values between successive scans. While the VVG board assigns one of 256 possible grey levels to a given pixel, it does not accurately distinguish 256 levels under most lighting conditions. A variability of 5–10 units should be allowed when calibrating grey levels so that pixels counted as 'background' in one scan do not suddenly count as 'subject' in the next, giving a false or inaccurate location. Even with proper calibration, the location recorded for a stationary subject may vary a few pixels in any direction. Effects of this variability may be minimized by proper setting of the 'threshold distance' in the data summary menu.

The relatively slow speed of image digitization lends some artificiality to the data, as do some heuristic components of algorithms designed to overcome this limitation. The motion of the subject relative to the scan may cause elongation or compression of its image, leading to possible inaccuracies in the location recorded. Using a smaller window size and a lower scanning resolution helps to overcome this problem by allowing a shorter scanning interval. As those factors are reduced, however, the program's ability to reliably track the subject becomes more limited.

The median location recorded for the subject is not the most appropriate in terms of sensory organs which are mostly located on the head. This discrepancy would be negligible if the subject were small relative to the observation area. Locations recorded for small subjects are also less affected by changes of body shape and orientation. Possible algorithms to find the anterior end for certain body shapes were discarded because of their heuristic nature, high complexity, and limitations in processing speed and image resolution.

Bias and loss of precision are possible when the program 'loses track' of the subject, especially for statistics based on distance travelled and turns. The program could lose track if the subject moves rapidly out of the scanning window or if it enters a shelter. In both cases, the program may be set to scan the full observation area until the subject is 'found'. However, as a full scan takes longer than a regular window scan, slight irregularities in the equality of time intervals between data records may occur. This is because a data record cannot be written until a given scan is complete. The effect of these irregularities may be minimized by increasing the time intervals between data records. The presence of a shelter may add further bias as the subject is assumed to be stationary when underneath it.

There are two main considerations on the adequacy of calculated statistics to represent the actual locomotor behavior of the subject. The first is that data collection is time-based rather than event-based. Observations collected by an event-based system such as a photocell array can be directly correlated to a movement by the subject, assuming the subjects cannot pass undetected between photocells. The tracking system described here, however, samples the instantaneous location of the subject at fixed time intervals. Accordingly, statistics such as number of trips to a given portion of the observation area or number of turns should be more correctly referred to as indices of the actual number of trips or turns. The subject could in actuality make several trips or turns between observations. The concept of a turn is artificial under any sampling regime as criteria defining it must be decided somewhat arbitrarily.

The second consideration on the adequacy of calculated statistics concerns the relative precision of the various locomotor statistics for a given sampling frequency (i.e. time interval between scans and data records). Indiscriminate use of all possible statistics is not appropriate, even though all are calculated. For example, a given sampling frequency may give adequate precision for the number of observations in defined portions of the observation area. The same sampling frequency may, however, be insufficient for a detailed analysis of turning behavior. Conversely, a higher sampling frequency may be adequate for detecting subtleties in turning behavior but, for the same duration of tracking, use up much more disk space and generate a lot of redundancy in statistics based on numbers of observations.

In summary, users should not place blind faith in the precision and accuracy of the tracking system. The limitations of the system force careful consideration in selection of options and interpretation of results.

ACKNOWLEDGEMENTS

R. Creedy and D. Genner of Computing Support Services, Department of Computing and Information Sciences gave some invaluable tips on Assembly language programming. This work was supported by a Natural Sciences and Engineering

Council of Canada operating grant to JBS and an Ontario Graduate Scholarship to RMK.

REFERENCES

Korver, R. M. (1986) *Greater ecological relevance and computerization in assessments of pollutant avoidance in fish*. MSc Thesis, University of Guelph, Guelph, Ontario. 108pp.

Korver, R. M. and Sprague, J. B. (in press) Avoidance reactions of fathead minnows to zinc: laboratory tests with computerized tracking and greater ecological relevance. *Can. J. Fish. Aquat. Sci.*

12

Biomonitoring of polluted waters: three systems

John V. Martin
Transviron, Inc., Maryland, USA

12.1 INTRODUCTION

Biomonitoring is a biological signaling system in which the response of a living organism is used to detect the presence of some condition of interest or importance. Biomonitoring or bioassay has the advantage of being direct. If you are using a locally abundant fish species, for example, as a biomonitor for toxicity in a municipal or industrial effluent, and the fish are alive and well in a tank of the effluent, then you are fairly safe in assuming that effluent may be discharged to the environment without endangering that species of fish. Conversely, if those fish are dying within minutes of being placed into the tank of effluent and the dissolved oxygen is adequate, then there is reason to assume that discharge of that effluent will be hazardous to the health of the wild population of that fish species.

Because of this direct approach, a bioassay indicates what actually happens to the organisms in the environment we are attempting to protect. Chemical and physical measurements can sometimes be substituted for a bioassay but they can only predict or infer what will happen in the environment. For this reason biomonitoring remains the only direct, empirical method of assessing environmental effects.

Biomonitoring has been used in various forms for many years to signal the presence of some hazardous condition. Miners used to use canaries to detect the presence of toxic gasses in mines. In medical, pharmaceutical, and toxicological research, laboratory animals are used as biomonitors in the development of chemicals, drugs, or poisons such as pesticides. In recent years the toxicity of industrial and municipal effluent has been monitored with a variety of different organisms. Fish, shrimp, and *Daphnia* are commonly used in acute toxicity studies of industrial effluent. Photosynthetic capacity of phytoplankton has been used to assess the effects of passage through the cooling system of a power plant and respiration rate of a variety of organisms has been used to assess impacts of industrial and municipal effluents. It is respiration rate that is the subject of this paper. Specifically three related systems will be described that use respiration rate of activated sludge or fixed-film biomass organisms to detect the presence of toxic or inhibitory substances or to monitor the performance of a wastewater treatment plant.

The three systems that will be described are a large-volume laboratory respirometer manufactured by Techline Instruments (Fig. 12.1), an on-line respirometer

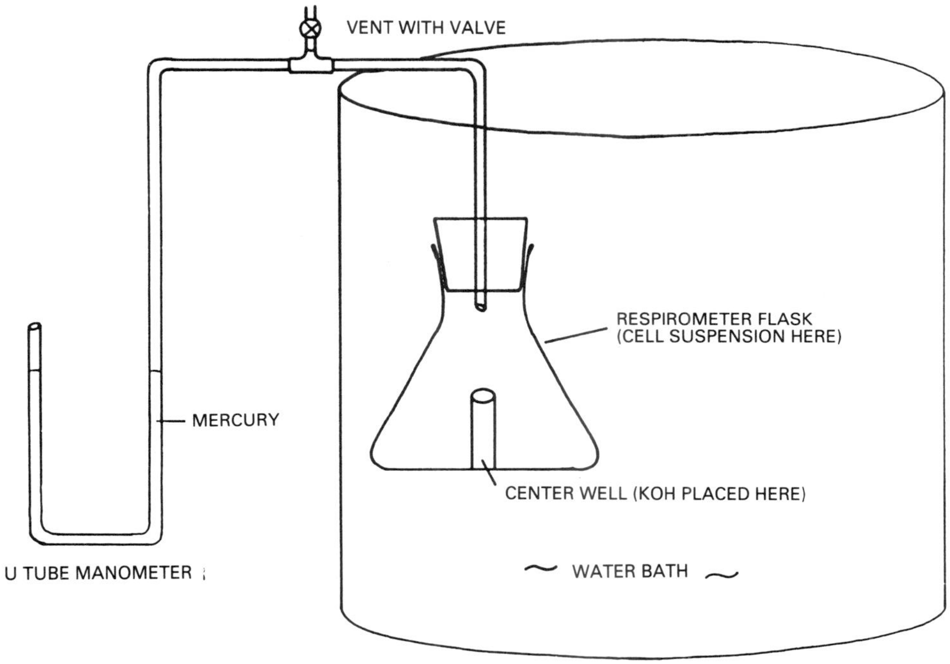

Fig. 12.1 — Warburg respirometer.

also manufactured by Techline Instruments (Fig. 12.2), and a toxicity monitor called Toxiguard manufactured by Eur-Control (Figs. 12.3 and 12.4).

12.2 LABORATORY RESPIROMETER

The laboratory and on-line respirometers both operate on the principle originally used by Warburg (Fig. 12.5). Carbon dioxide produced by aerobic respiration in a closed vessel is absorbed by strong caustic, usually potassium hydroxide, allowing oxygen consumption to be measured manometrically. Both the laboratory and on-line respirometers use this principle in measuring respiration rate, although in these respirometers the gas is actively recirculated (Fig. 12.6), while in the Warburg respirometer, carbon dioxide must diffuse passively into the potassium hydroxide scrubber.

At the Patapsco Wastewater Treatment Plant in Baltimore a procedure has been

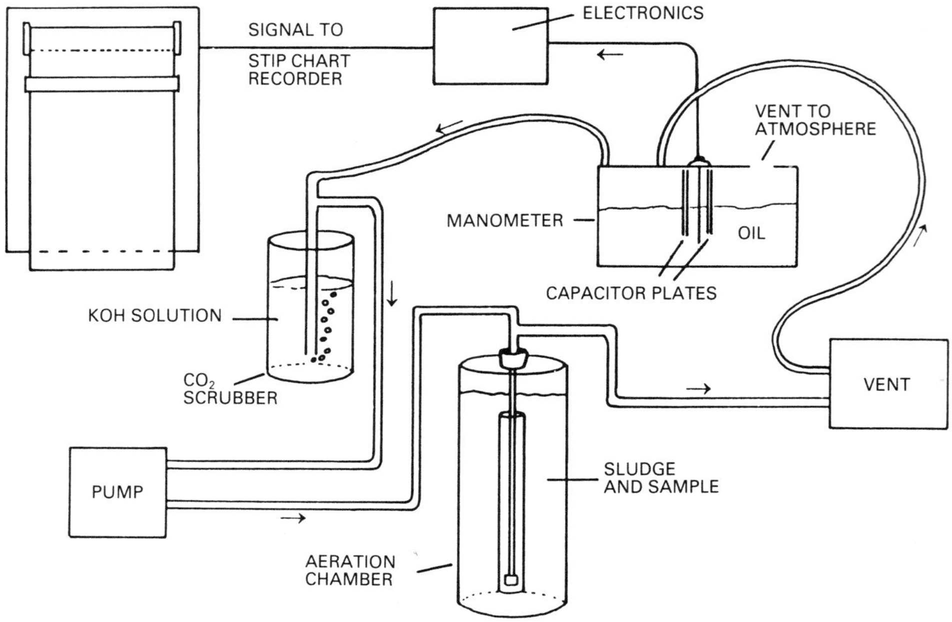

Fig. 12.2 — Techline respirometer.

developed for biomonitoring industrial and municipal wastewaters with the laboratory respirometer. The treatment plant's return activated sludge is used as the source of bioassay organisms. A sample of this sludge is diluted to approximately 0.5% solids and placed in the respirometer where the respiration rate under endogenous or starvation conditions and under conditions of excess substrate are determined. This latter is termed the blank rate and it functions as a basis of comparison for samples of industrial or municipal waste. Respiration rates are then determined for samples of treatment plant influent and industrial effluents under similar conditions of excess food and oxygen. If the respiration rate of sludge diluted with a sample is equal to or greater than the blank rate then that sample is judged treatable. If, however, the sample causes an inhibition of respiration rate, then a percentage inhibition is calculated and the information is used as an indicator or potential problems with the effluent from that source.

The respiration rate obtained from the respirometer in these determinations is an initial rate. The data are taken as soon as the chart recorder indicates less than 10% change in rate or slope in two consecutive five-minute intervals. In this way, a standard criterion is used for terminating a respirometer run that otherwise could continue for several hours, during which time the respiration rate would change gradually. The formula used to calculate percentage inhibition is the blank respiration rate minus the sample respiration rate divided by the blank respiration rate times 100, with all rates net or corrected for the endogenous respiration rate.

One phenomenon that is occasionally observed with this procedure is that an

Fig. 12.3.

industrial sample will exhibit a stimulatory effect resulting in a respiration rate greater than the 'best case' or blank rate. It has been suggested that the increase in oxygen uptake may be an artifact resulting from a portion of the bacterial biomass being lysed by the sample and providing additional food for the surviving cells. There may also be more complicated biochemical reasons such as uncoupling of oxidative phosphorylation causing the stimulation.

The laboratory respirometer can also be used for performing a short-term biological oxygen demand (BOD) determination or for any test involving biological consumption of oxygen. It also has almost limitless research applications.

12.3 ON-LINE RESPIROMETER

The on-line respirometer takes many components from the laboratory unit and adds a timer-controlled system that automatically drains and refills the aeration chamber,

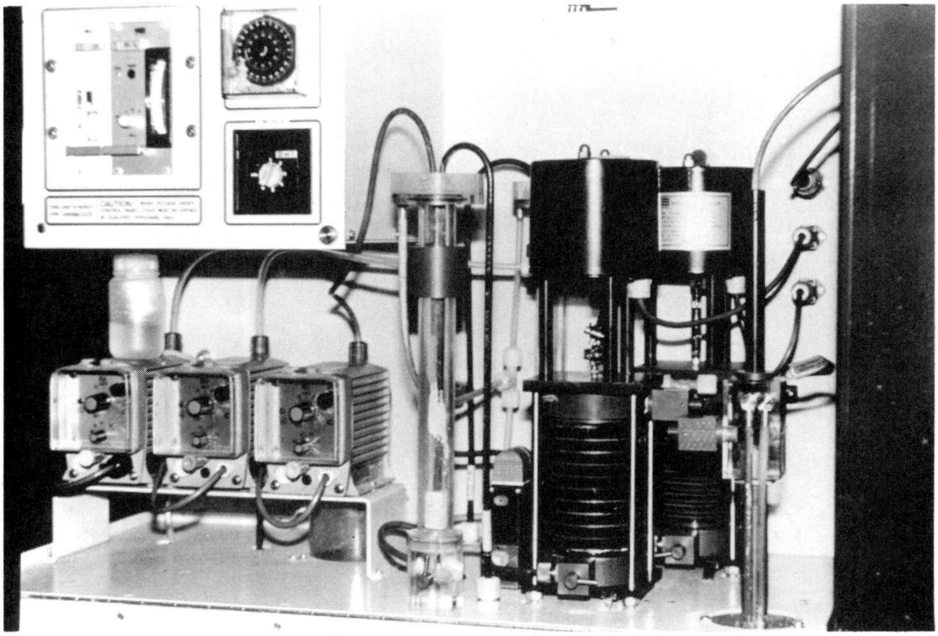

Fig. 12.4.

Fig. 12.5.

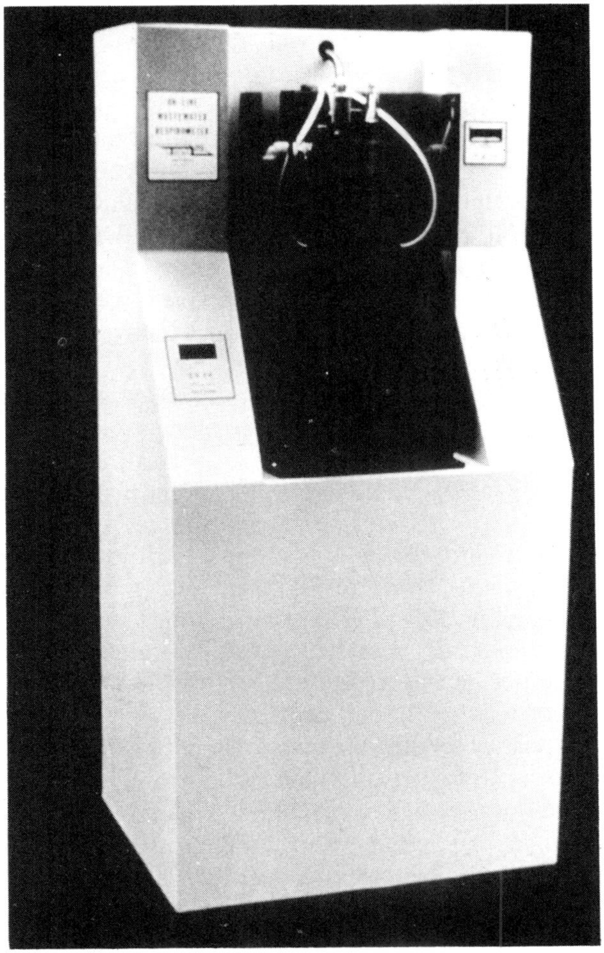

Fig. 12.6.

measures respiration rate and then repeats the cycle. The time is adjustable up to 99 hours, allowing flexibility in selecting the cycle time. Normally the instrument is set to cycle every 20–30 minutes. Additional flexibility is achieved in that the on-line unit can be connected to multiple streams, sequentially sampling each one, measuring respiration rate and moving on to the next. Although, theoretically, several streams could be monitored, in practice, the limit is three. Some examples of multiple-sampling-point installations are at the inlet, mid-point and outlet of an aeration tank, at the outlets of three different aeration tanks, at the influent, primary effluent and secondary effluent of a biological treatment plant, or at up to three different streams at an industrial facility such as a chemical plant.

As the respirometer progresses through its sequence, the rates of respiration of activated sludge are measured and displayed as the slope of the line on the instrument's chart recorder. A second pen on the recorder is arranged to produce a trend line that connects the tops of each subsequent peak produced as the respirometer cycles. The trend line is useful in detecting slow changes that would not be noticed in a short term observation. An example of using the data from an on-line unit for process control is if the three sampling points are at three locations within an aeration tank, and the respiration rates of the mid-point and outlet end of the tank are nearly the same; this indicates the tank is underloaded. On the other hand, if the respiration rate of the outlet is not much different than at the inlet, then this indicates that the tank is not removing much BOD. This could indicate that the tank is overloaded or that the sludge is not functioning efficiently, perhaps due to a toxin in the water or inadequate aeration. In any case, the data from the on-line unit can be used to provide a real-time indication of the efficiency of an aeration tank. Decisions on whether to place additional units in service, change wasting rate or aeration rate can all be aided by data from an on-line respirometer. Installations that sequentially sample different streams at a manufacturing facility can provide important information to a plant manager regarding the toxicity of his plant's effluent to activated sludge organisms.

On-line respirometer installation in aeration tanks or on other mixed-liquor or sludge-containing streams need only to pump samples into the aeration chamber and measure the respiration rate. All other installations must be provided with a separate source of seed. If a biological treatment works is located nearby, sludge may be obtained from this source and used for seeding the respirometer. If no source of seed material is available nearby, sludge organisms may be cultured in continuous reactors for the purpose of seeding the respirometer.

Data obtained from a properly installed and operated on-line respirometer can be used to detect the presence of toxic or inhibitory substances and to observe long-term trends in treatment plant performance. Its ability to simultaneously monitor up to three streams increases its usefulness to both treatment plant and industrial plant operators.

12.4 TOXIGUARD

Toxiguard is a system best suited for industrial effluent or treatment-plant influent monitoring. Respiration rate of a bacterial biomass is still the indicating parameter, but in this case it is a fixed-film biomass and instead of measuring respiration rate by manometrically measuring changes in gas volume, the Toxiguard continuously measures the dissolved oxygen concentration in the effluent stream from the fixed film biofilter (Fig. 12.7). Effluent enters the unit at an overflow chamber and is pumped into the aerator. Synthetic sewage is also pumped into the aerator. The mix of synthetic sewage and effluent sample is aerated until the dissolved oxygen is approximately 7–8 mg/l. This mix is then fed to the biofilter where the organic matter in the sample is broken down and oxygen is consumed by the biomass. Under normal circumstances, when the biomass is healthy and the feed does not contain substances that inhibit respiration, the dissolved oxygen concentration in the biofilter effluent will be nearly zero. If some inhibitory substance poisons the biofilter, its capacity to consume oxygen will be reduced and the dissolved oxygen in the effluent stream will start to rise. A probe continuously monitors the dissolved oxygen in the biofilter

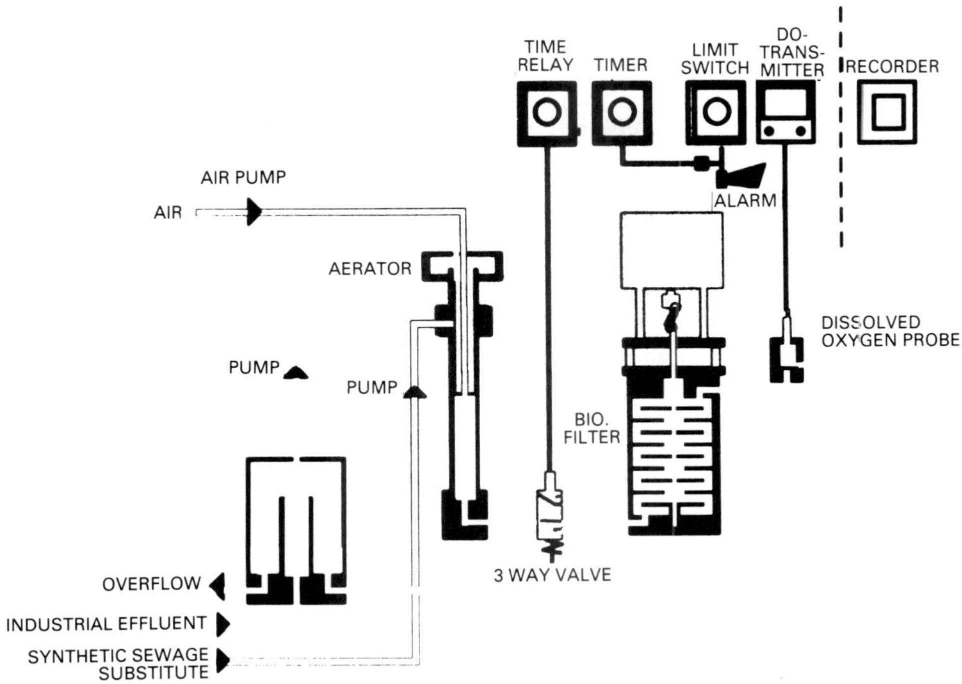

Fig. 12.7.

effluent and if it rises above a certain present level, an alarm sounds alerting the operator of possible toxic inhibition in the influent wastewater.

The oxygen probe must be calibrated occasionally but the Toxiguard has an internal calibration system that the unit goes through automatically on an interval selectable from once per hour to once per 24 hours. Normally, internal calibration is required only once per 24 hours. The calibration system uses a three-way valve to divert the sample around the biofilter and directly to the oxygen probe. The alarm function is disconnected during this sequence. When the oxygenated feed reaches the probe, the recorder indicates the dissolved oxygen concentration of the biofilter influent. This should be approximately 7–8 mg/l. After approximately three minutes, the three-way valve returns the flow to the biofilter. Another minute elapses and then the alarm function is reconnected. If this internal calibration does not indicate a dissolved oxygen concentration of 7–8 mg/l in the feed to the biofilter, the cause should be investigated. The oxygen probe may be out of calibration, or the air pump or one or both of the sewage feed pumps may not be functioning properly.

If a reliable flow of wastewater to the unit can be arranged, the Toxiguard can be an effective means for timely detection of the presence of substances in that stream that will upset a wastewater treatment plant.

12.5 COMPARISON OF FEATURES

12.5.1 Installation and physical requirements

The laboratory respirometer requires 115V A.C. power only. No permanent plumbing or other installation is required. A sink with hot and cold running water is useful and good temperature control in the room where the unit is located will improve results. The unit occupies approximately six lineal feet of bench space, and drawer or cabinet space for storage of supplies and accessories is also useful. The on-line respirometer requires a pipe or hose to each stream being monitored. Multiple stream installations must also be equipped with a multiport valve so that the different streams being monitored can be introduced into the respirometer sequentially. A supply of cold water is required for chamber rinsing between cycles and results will be more consistent if the instrument can be installed in temperature-controlled space. The Toxiguard requires 115V A.C. power and permanent plumbing to the stream being monitored. As with the respirometers, temperature-controlled space will improve results. At a minimum, the unit must be protected from freezing. A sink with hot and cold running water and a supply of synthetic sewage are needed to complete the installation. Both the Toxiguard and on-line respirometer require approximately 16 square feet of floor space, allowing access to the sides and back of the cabinet.

12.5.2 Timeliness of results

The laboratory respirometer is capable of producing results of a respiration rate determination in approximately 30 minutes. Allowing time for calibration of the instrument and determination of endogenous and blank respiration rates, data from the first sample will probably be available in two hours or less. Subsequent samples will take longer. If the respirometer is being used for BOD determinations, it will take at least two hours and probably more before results are available. The on-line respirometer can produce a new respiration rate curve every 20–30 minutes, so if the unit is installed to monitor a single stream, results can be as current as a half-hour or less. If the on-line unit is monitoring three streams, and is cycling every half-hour, the data will be no more than 90 minutes old. The Toxiguard is capable of responding to a toxic event in 6–10 minutes. This is the fastest response time of all three systems and if operators react quickly, may be timely enough to prevent major damage to a treatment plant's biomass. The Toxiguard is equipped with a back-up filter so that, if one becomes poisoned, a healthy filter will always be available for use as soon as the spill has passed. The poisoned filter can be allowed to recover fully: a process that takes from one to 12 hours.

12.5.3 Accuracy of results

The laboratory respirometer can be run with an accompanying control on every determination adding strength and validity to the data. The on-line produces data that are best interpreted on a relative basis, whereas the Toxiguard produces very accurate data but only a yes or a no. Toxiguard does not provide data on degree of inhibition.

12.5.4 System sensitivity

The sensitivity of all three of these instruments is a function of the sensitivity of the organisms to the particular toxin or inhibitory substance in question. With the laboratory respirometer, however, the degree of toxicity can be assessed because a parallel control can be run to provide a basis for comparison with every determination. With the on-line unit, it is only possible to state that a waste stream is more or less inhibitory than the previous sample or samples from earlier that day. The Toxiguard is highly sensitive to any substance that inhibits the ability of the organisms in the biofilter to absorb oxygen but one cannot reliably assess the degree of toxicity with data from the Toxiguard.

12.6 CONCLUSION

Three different biomonitoring systems in use at the Patapsco WWTP have been briefly described. The laboratory respirometer is an accurate and flexible tool allowing determinations of toxicity and BOD to be done in a couple of hours. It also can be used as an effective research tool. The on-line respirometer provides nearly continuous monitoring of up to three streams. Monitoring most streams except mixed-liquor or other sludge-containing streams requires a separate source of seed. The Toxiguard provides the fastest response to toxic or inhibitory conditions, but is the least flexible in its applications.

13

Development of a system for continuous biomonitoring of a domestic water source for early warning of contaminants

Edmund H. Smith and **Howard C. Bailey**
ANATEC Laboratories, Inc., 435 Tesconi Circle, Santa Rosa, California 95401, USA

13.1 INTRODUCTION

The Russian River basin, located about 75 miles north of San Francisco, California, supports several important beneficial uses, the most critical is its use as a drinking water source for nearly 450 000 people. The municipal water systems provide limited, if any, off-stream storage and only minimal water treatment by chlorination (Klamt, 1985). Many of these municipal diversions are Ranney collectors placed in or adjacent to the river, drawing upon the river underflow through collection systems in the alluvial gravels. A recent spill of 30 000 gallons of formaldehyde into the river caused great concern to water purveyors because of the fast, direct input of the contaminant into the supply systems and potential long-term contamination of alluvial gravels. The need for some type of continuous early warning was clearly evident. If the water purveyors could receive a warning in time to determine the severity of the contamination and have time to cease pumping before the supply system and stream-bed gravels were contaminated, the public could be better protected.

Because of this spill incident and the fact that other contaminants of concern have been found in the river during routine chemical monitoring, the California North Coast Regional Water Quality Control Board sought and received Federal 205(J) funding to test the feasibility of an early warning system in the Russian River. A team of agencies and private corporations was assembled to undertake the development of the system. The team consisted of staff from the Regional Board, Sonoma Water Agency, Motion Analysis Corporation, and ANATEC Laboratories. There was funding for the project from 205(J) sources, and the Regional Water Quality Control

Board, corporate contributions from the two private firms and support from the Sonoma Water Agency.

After investigating a number of possible approaches to continuous monitoring of the river it was decided to develop and test new monitoring techniques. The conclusion was to combine continuous behavioral monitoring with composite resin-column sampling as a chemical check on the behavioral response of the test organisms. In this way, the behavioral response of the organisms could be used to detect contaminants and the causative agents determined by subsequent chemical analysis. This approach is similar to that described by Gruber *et al.* (1981).

Using aquatic organisms for continuous toxicity monitoring is not a new concept. A detailed review of early warning systems employing biological monitoring was presented by Cairns and van der Shalie (1980). In the 1960s, fish survival was monitored in flowing wastewater (Jackson and Brungs, 1966), and long-term monitoring of fish activity has been utilized in Sweden (Hasselrot, 1975). Biological monitoring systems employing rheotaxis of fish were used in Germany to determine if the fish could maintain their position in a moderate current after or during exposure to a toxicant (Besch *et al.*, 1977). With the advent of computers and closed-circuit video systems, new possibilities developed for continuous biomonitoring (Gruber *et al.*, 1981; Lubinski *et al.*, 1977). A number of early warning systems are now on the market in Europe and are being deployed in the field. The Grieder system produced by the Endeas and Hauser Instrument Company uses a change in the pulse frequency from a Nile River fish *Gnathonemus petersi* as an alert to chemical exposure. This type of fish emits pulses at approximately 18 Hz. Upon the concentration of a toxic substance in the water reaching a certain level, the activity of the fish decreases, coupled with a marked reduction in the pulse rate (Ewen, 1987). Aquatest is a continuous monitoring system manufactured by Quantum Science, Ltd based upon the disorder of the positive rheotactic response of fish to a chemical (swimming against a current; Anonymous, 1986). This system is designed to respond to accute levels of contaminants. The WRc Mk III Fish Monitor uses the ventilation frequencies of several individual trout which are measured separately and compared with historic norms. The monitor is deemed to have responded when current frequencies are significantly different from those measured between one and two hours earlier (Evans *et al.*, 1986). With this existing data base of computer-based biomonitoring, the basic approach selected for the biological monitoring aspect of this project was to use a modified commercially available (Motion Analysis Corporation) system which the private firms involved in the project were familiar with. This computer-controlled, video-based automated system was first described by Greaves and Wilson (1980) as a research tool which followed the actual behavioral patterns of the test organisms. The basic system was further modified for application to continuous monitoring of river water.

13.2 METHODS AND MATERIALS

13.2.1 Selection of the test species

A set of criteria were developed for species selection which would be applicable to the computer-based monitoring system developed for this project. These criteria were as follows:

ease of organism acquisition and maintenance,
sensitivity to a wide range of toxicants,
organisms that are in constant motion,
suitability for use in flow-through systems,
the organism should be small to reduce the size of the exposure chamber and perhaps increase sensitivity,
availability of sufficient toxicological data to assess possible response thresholds (behavioral or LC_{50} data), and
the organism can tolerate the ambient water quality conditions of river water (daily and seasonal variations in temperature, dissolved oxygen and turbidity).

The above criteria were applied to a number of standard bioassay test organisms. The first three selected for preliminary testing were steelhead and rainbow trout (*Salmo gairdneri*), Chinook salmon (*Oncorhynchus tshawytscha*) and *Daphnia magna*. The selection of both fish and an invertebrate was done based on the theory of widening the range of behavioral response to toxicants and to increase the sensitivity to a greater array of contaminants.

Steelhead were obtained from the Warm Springs Hatchery located on a tributary of the Russian River. Rainbow trout were purchased from Thomas Fish Company in Novato, California. The Chinook salmon were obtained from the Nimbus Fish Hatchery on the American River. The size range of the steelhead and rainbow trout was between 35–65 mm and for the salmon 30–40 mm. The fish were transferred from the hatchery to recirculating systems (Living Stream; Fridged) at ANATEC Laboratories and maintained for one month before testing. Temperature was held constant at 12°C. *Daphnia magna* stock were obtained from SRI (Stanford Research Institute) and cultured in our laboratory. Temperature was maintained at 22°C.

13.2.2 Flow-through bioassay system design

13.2.2.1 Fish test system design

The test chamber flow diagram (Fig. 13.1) illustrates how the system is configured. During the early stages of testing, the control water was dechlorinated city water (carbon filtration); later, well water was used. Test toxicants were delivered by metered pump. Control and toxicant could be supplied down either channel of the test chamber depending upon behavioral response and to verify avoidance–preference behavior.

The chamber for the fish avoidance–preference testing was modified from the system developed by Westlake and Lubinski (1976). The overall size of the chamber is 91.4 cm long, 48.3 cm wide and 7.6 cm deep. The chamber is made of glass and is divided into three compartments: control, test and avoidance–preference (Fig. 13.2). Test fish are placed in both control and test chambers. The flow is channeled to both chambers and the toxicant discharged into the appropriate channel. After the water has flowed through the two divisions, it enters an avoidance–preference area where the fish have a choice to either avoid or be attracted to the test material, or not to change their response pattern from the control conditions. The glass partitions

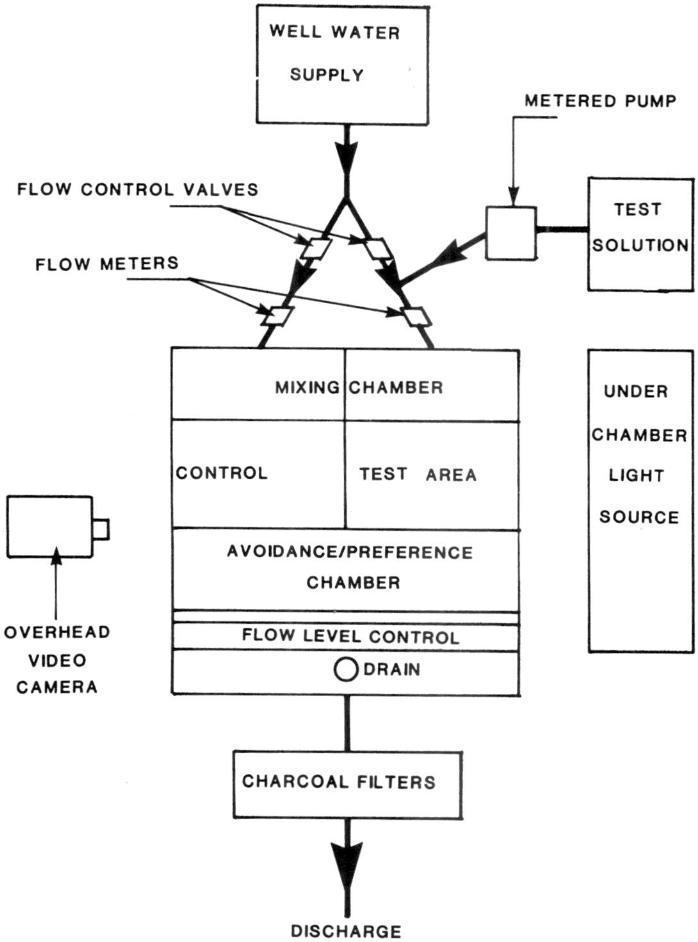

Fig. 13.1 — Diagram of the fish behavioral testing system used in the laboratory.

between the different areas were sandblasted so that the fish in each area could not see the behavior of neighboring fish or their mirror image.

For the fish avoidance–preference tests, the decision was made to employ a steep gradient rather than a shallow gradient between test and control solutions. Since a single test concentration was used for each test, the steep gradient was selected in order to provide the best opportunity for the sense organs of the fish to discriminate between the opposing bodies of water (Larrick *et al.*, 1978). These gradients provide the necessary cues for successful directed movements such klinotaxes and tropotaxes. In addition, kinesis is apt to be more successfully executed in very steep gradients. Dye tracer and fluorometric tests were conducted to determine the time necessary for mixing, when test chambers reach maximum concentration and the stability and sharpness of demarcation between test and control water masses within

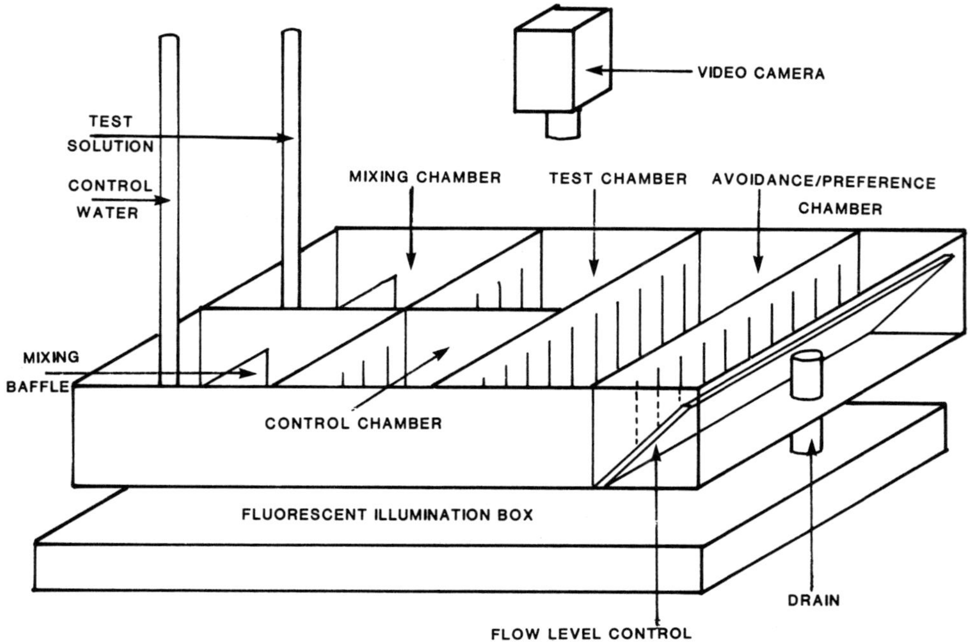

Fig. 13.2 — Diagram of the avoidance–preference chamber used for fish behavioral testing.

the avoidance–preference chamber. Based upon the dye tests it was determined that the zone of dilution on either side of the theoretical center division of the water mass was 3–4 cm wide (Fig. 13.3).

Calibrated flow meters were used to supply the test chambers with water. The best demarcation between water masses was obtained at 4 l/min through each chamber for a total flow of 8 l/min. Ten fish were placed in each of the divided areas for a total of 30 organisms per test. The chamber was illuminated with red-filtered, diffused fluorescent light from below throughout the test period.

13.2.2.2 **Daphnia** ***test system design***

The *Daphnia* test system consisted of a control water source (as described earlier), supplied to a gravity tank (Fig. 13.4) in order to provide constant head pressure. This dilution water was allowed to flow freely at a rate of 60 ml/min into a funnel which served to break any siphoning effects. This flow rate was selected because it was not strong enough to effect the swimming behavior of *Daphnia* and still allowed relatively rapid turnover of the chamber (approximately one volume change every seven minutes). To the bottom of the funnel was attached a line feeding into the *Daphnia* chamber. A stock phenol solution was stirred continuously with a magnetic stirrer and was dripped into the funnel at a predetermined rate from either a metered pump or Marriotte bottle. The algae *Selenastrum capricornutum* was delivered to the funnel in the same manner as the test solution. Algal solution at a concentration

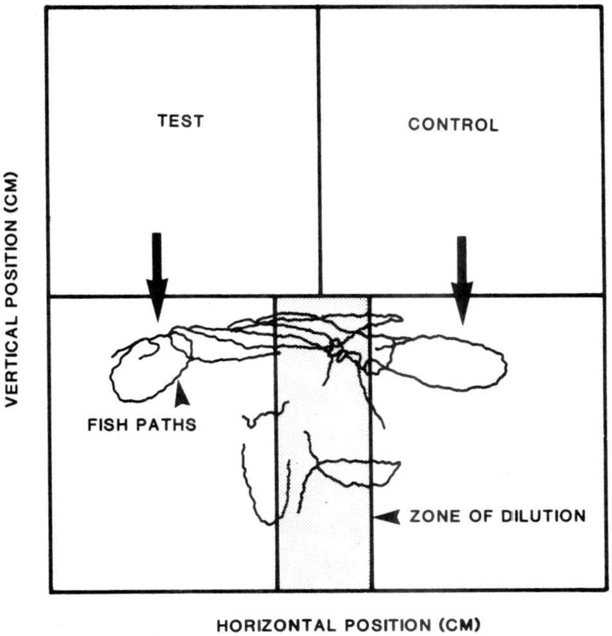

Fig. 13.3 — Diagram of the mixing zone in the avoidance–preference chamber with 15 s of path data plotted.

of 33 000 cells/ml was delivered to the funnel at a rate of 1 ml/min to achieve a final concentration of approximately 500 cells/ml in the chamber. The *Daphnia* were fed continuously during the experiment.

The chamber designed for the *Daphni* testing consisted of a 12.7×17.8-cm plastic aquarium 1.9 cm deep. The sides of the unit (Fig. 13.5) were drilled lengthwise with supply and drain holes 9.5 mm in diameter and numerous 0.5 mm holes were placed along both sides of the aquaria to allow the water to enter and leave the chamber. The top of the unit was sealed and the entry plug was placed near the bottom of the chamber. Two Teflon fittings were threaded into the top at each end to deliver and remove the water from the tanks. A third Teflon fitting allowed the air to escape when the chamber was filled with water. Twenty adult *Daphnia* were placed in the chamber.

The chamber was illuminated from behind by an infrared light source with an emission spectrum of 820 nm. At this wavelength, the *Daphnia* are not sensitive to the light source. A fiber-optic, low-intensity polychromatic light was placed directly above the aquaria to keep the *Daphnia* moving and off the bottom.

13.2.3 Microcomputer-based video system

A standard CCD video camera (sensitive to infrared light) was used to capture the images of the test organisms in the two systems described above. For the fish chamber, the camera was placed directly above the unit with the fish illumin-

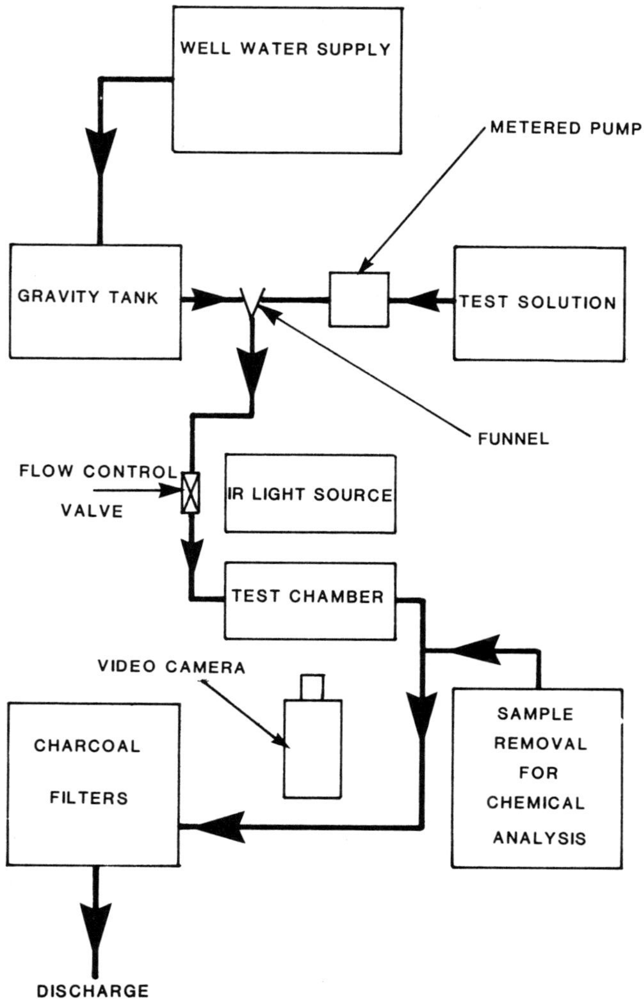

Fig. 13.4 — Diagram of the *Daphnia* behavioral testing system used in the laboratory.

ated from below. This lighting arrangement created strong contrast between the organisms and the background. The *Daphnia* chamber camera was placed directly in front of the tanks with the infrared light unit providing the main back lighting source for illuminating the outlines of the *Daphnia*.

The analog video images were fed into a proprietary video processor that digitized the image outline and input the data into a microcomputer-based graphics workstation (Sun or IBM). The raw or processed images can be displayed on a monitor, giving the operator immediate visual feedback.

Once the digitized image is stored on the workstation's hard disk, the software manually or automatically transforms these raw data into graphic images that depict

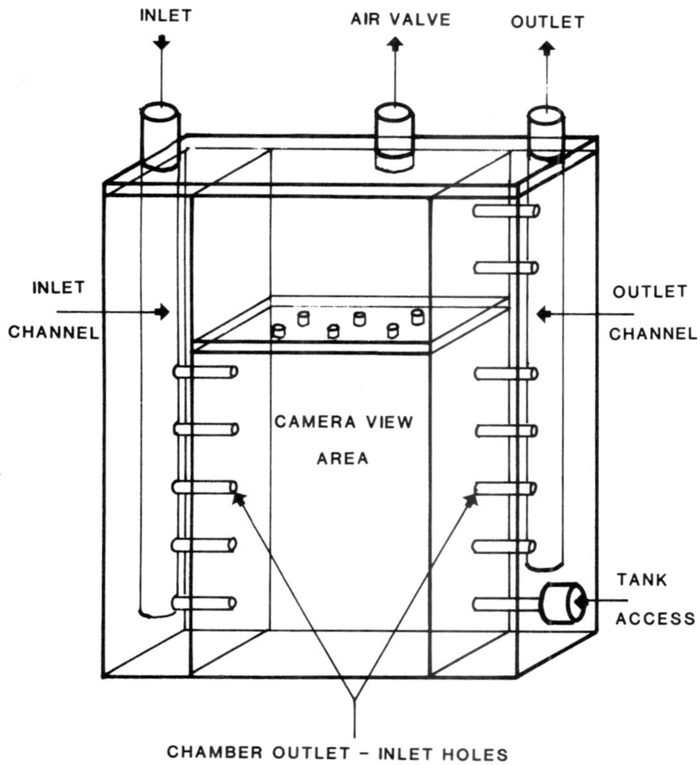

Fig. 13.5 — Diagram of the *Daphnia* test chamber.

the path of the motion. Once the paths have been designated, other parameters of motion can be calculated (linear and angular velocity, X–Y coordinates, change of direction and others). The calculations can be displayed as alphanumeric tabulated summaries and in graphic form (line graphs, bar charts, three-dimensional images). The results may also be merged into an existing database for accumulating behavior summaries. Hard copy of all screen displays can be generated on the system's printer/plotter. After many behavior experiments, two indices of movement seemed to be the best for the automated system; these were linear velocity and location (in both the X and Y coordinates).

13.2.4 Behavioral observations and data analysis

After the fish were acclimated in the recirculation system for four weeks, 30 individuals were transferred to the preference–avoidance chambers for acclimation. The fish were acclimated for about three days in the chamber as determined by 24-hour automatic data capture (15 seconds of behavioral data gathered every seven minutes). When the behavior was stabilized, as determined by linear velocity, the entire data set was checked statistically to establish a control-behavior baseline. The same procedure was used for *Daphnia,* although the acclimation period was shorter (24 hours).

During the laboratory test period, the behavioral response was captured on video tape rather than directly on the hard disk of the computer. This was done so that the tape could be scanned visually and so that the test could be re-analyzed using different time sequences or different measurements of motion.

One hour of control-behavior data was captured on the video tape before the test material was added to the chamber. After the test material began entering the chamber, another hour of data was gathered during the exposure and for an additional half an hour after the chemical inflow ceased and was replaced by control water. Statistical analysis was conducted on the control data which was compared with the three-day 'norm'. If there was no significant difference, the test response was compared statistically with the norm. If there was a significant difference between the test control period and long-term norm, the test was repeated.

13.2.5 Early-warning remote test site

A site was selected along the Russian River which would afford easy access and protection from vandals, and would be located above the winter flood level. The platform on top of a Ranney collector operated by the Sonoma Water Agency met the criteria. A small building was constructed to house the exposure chambers, computer system and resin-core sampling equipment.

A block diagram (Fig. 13.6) illustrates how the early warning system is configured. Water is supplied from the river by a submerged stainless-steel pump encased in a PVC 15.2-cm pipe, perforated with a series of small water-access holes and anchored to a concrete fish ladder. A length of 1.3-cm stainless-steel tubing carries the water to the platform 24.6 m above the river bed. The flow is 17.1 l/min at the platform.

At this time the *Daphnia* test system is the only one in operation at the remote test site although provision for the addition of the fish has been made.

Two types of flow-through systems are housed in the enclosure on the platform. One system is a closed-type, flow-through unit which acts as a control. The water is recirculated through a biofilter and ultraviolet sterilizing light into the fish and *Daphnia* test chambers and returned back through the biofilter. Reconstituted deionized water is adjusted to match river water quality (hardness, pH, alkalinity, temperature). Flow rates are 60 ml/min for the *Daphnia* tank and 4 l/min for the fish chamber. The fish chamber does not have the avoidance–preference separations used in the laboratory. This change was done because there was no constant supply of control water. Positional data will still provide information on avoidance–preference response of the test fish.

River water is supplied to the exposure chambers through a large (38.1×122 cm) PVC Venturi filter to remove larger particles. Excess water is discharged, and the water used for the fish and *Daphnia* chambers is filtered through a 30-μ and 5-μ filter to avoid particle build-up over time. The water is then passed through stainless-steel tubing immersed in a temperature-control water-bath held at 22°C. Flow rates are the same as the control system.

Daphnia are fed from a container which provides a continuous supply of the algae *Selanastrum capricornutum* to the chambers. Nutritional sources should be available during *Daphnia* testing (Barera and Adams, 1983) and algae seems to supply the best nutritional value (Buikema, *et al.* 1980). Fish will be fed by a continuous auto-feeder.

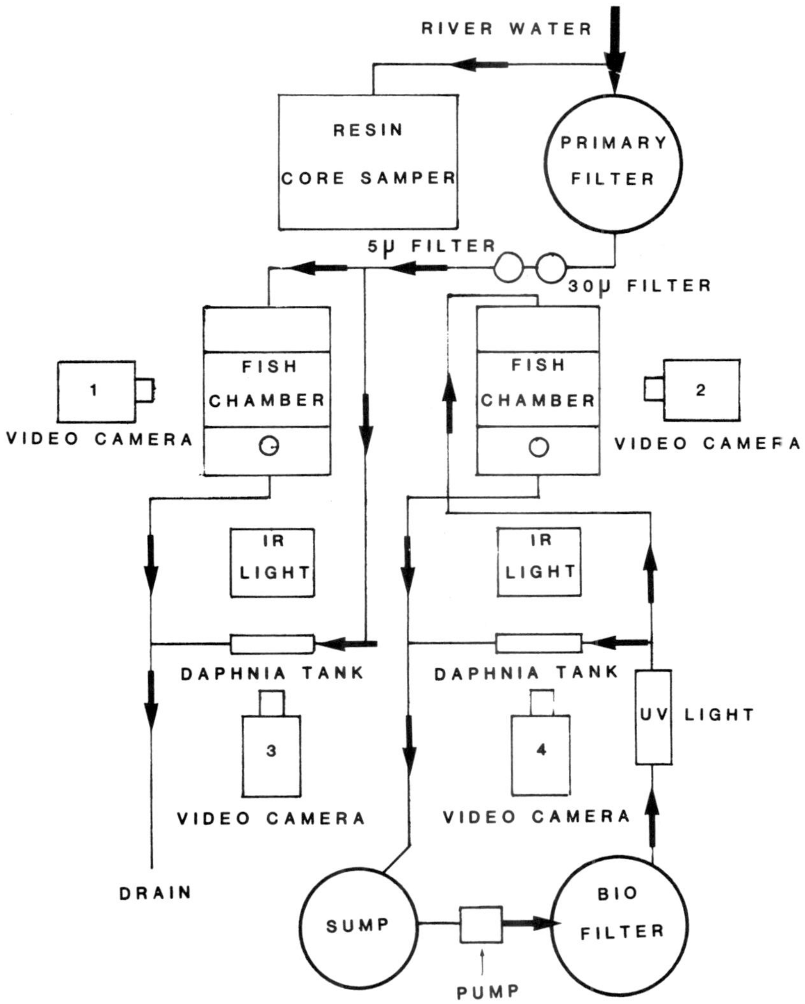

Fig. 13.6 — Diagram of the fish and *Daphnia* test system used at the field site.

Filters were added to the system after determining that the toxicants of concern for this project were not likely to be closely bound to the fine slits in suspension during high-flow periods. In addition, the Water Agency was not concerned with particle-born toxicants since the large, natural filter beds would remove them before the water entered the distribution system.

The computer interface with the test chamber is illustrated in a block diagram (Fig. 13.7). Water is supplied to the test environment and the behavioral response is followed with four video camera. Fifteen seconds of video data are captured from each camera sequentially. The analog video images are sent to the video processor where the image can be viewed on the monitor and/or recorded on a video cassette recorder. The processed video information is sent to the host computer (IBM-AT

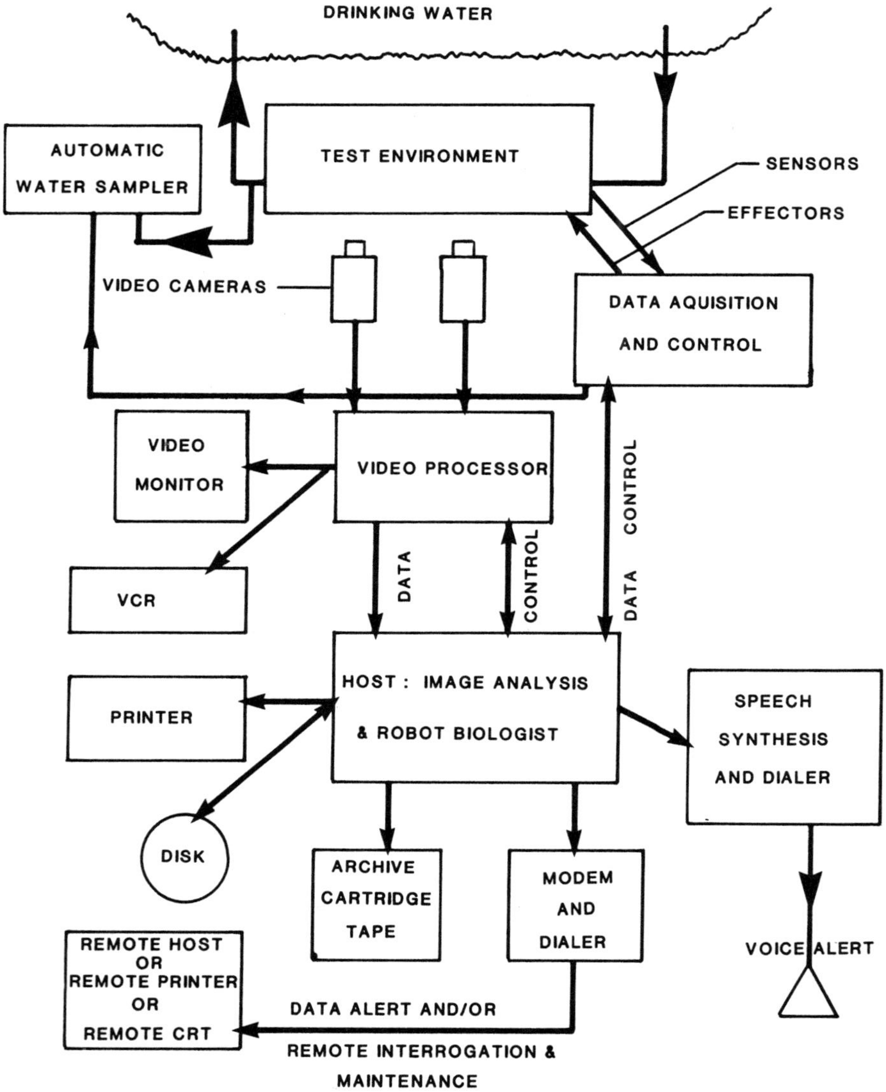

Fig. 13.7 — Flow diagram of the early warning system interface with the computer, video processors and alarm component.

72 Mb, hard disk) where the behavioral information is summarized from each camera in separate files. Each set of data is compared statistically to the 'normal' behavioral patterns to detect change. The analyzed data can be distributed to disk, cartridge tape or sent by modem to a host computer at a different location. If the statistical analysis shows significant variation from the normal behavior and the results satisfy the alert warning criteria, a speech synthesiser and dialer can activate a voice alert or interact directly with the water purveyor's system (turn off valves or pumps).

13.2.6 Water quality and temperature control

Water quality measurements in the laboratory were taken daily from the supply-water and organism-holding containers. These measurements included temperature, pH, dissolved oxygen, ammonia, alkalinity, hardness and conductivity. During the test periods, the dilution water was matched to the holding water values of hardness at 120 mg/l and alkalinity at 130 mg/l (both as CCO_3). Temperature was controlled in the recirculation system for the salmonid tests at 12°C and the room temperature was also held at the same temperature. *Daphnia* control water temperature was controlled by holding the room temperature constant at 22°C.

At the field site the measurements listed above will be monitored by probes and the data evaluated by the computer through a Hewlett-Packard data logger as part of the early warning system water-quality evaluation procedure. These data will be evaluated along with the behavioral information to form part of the alarm system. During test periods, samples of the toxicant were removed from the discharge side of both the fish and *Daphnia* chambers and analyzed for phenol using the absorptiometric method (Ochysnski, 1960). The outflow sampling station was located outside of the test room to avoid disturbance during the test period. Phenol samples for the *Daphnia* were taken every 30 seconds from time zero to eight minutes. From eight minutes to 10 minutes, samples were taken every minute, and from 10 minutes to 30 minutes samples were taken every five minutes. Additional samples were taken at 60 minutes, 120 minutes and at 22 hours. Effluent was collected for 20 seconds, resulting in sample volumes of approximately 20 ml. Fish sampling periods were established at every minute during the first 15 minutes of the test and then every five minutes for the remainder of the experiment.

13.3 RESULTS

13.3.1 Prototype laboratory testing

The laboratory testing protocol was developed to satisfy two major goals: to test the behavioral response to known toxicants and to test the prototype computer system that will operate at the remote site. The results are preliminary with emphasis placed upon the *Daphnia* experiments. Tests on fish were designed simply to determine if behavioral responses to a known toxicant could be detected by the prototype early warning system.

The first set of toxicants considered for testing salmonids and *Daphnia* included those that have been historically found in the Russian River (formaldehyde, DDT (dichlorodiphenyltrichloroethane), pentachlorophenol, phenol). Phenol was selected for the first test because it has been used as a reference toxicant (Lewis, 1981; Klaverkamp *et al.*, 1985) and because there is a good data base on phenolic compounds. The results from the phenol testing will be discussed as an indication of the response time and level of behavioral activity induced by exposure to the toxicant as recorded by the computer system.

The parameters selected for analysis of behavior follow the ones suggested by Giattini *et al.* (1982), as well as some additional measurements made possible with the new computer and software:

Linear velocity, which is a measurement of the speed of the organism through time.
X-coordinate, which is a measurement of the position of the organism on the X-axis through time.
Y-coordinate, which is a measurement of the position of the organism on the Y-axis through time.
The rate of change of direction. In the analysis of translational movement, rate of change of direction is the absolute value of the (translational) angular velocity. Angular velocity is the first derivative of continuous direction of travel with respect to time. Thus, it is defined for every point or instant in an object's path and is expressed in units of degrees per second. The rate of change of direction is important as an index of turning regardless of the direction of the turns.

The data are time-dependent and therefore represent behavioral response through time. This same approach was used by Ishio (1964) to determine concentration by plotting positions against toxicant level.

13.3.2 Salmonid response to phenol

Preliminary tests on salmonids were conducted using phenol at a concentration of 12.0 mg/l which is higher than the reported LC_{50} value for rainbow trout (8.9 mg/l; DeGraeve *et al.*, 1980). Based upon the analytical results, a concentration curve was developed for each test sequence (Fig. 13.8). Each dot represents the mean behavior

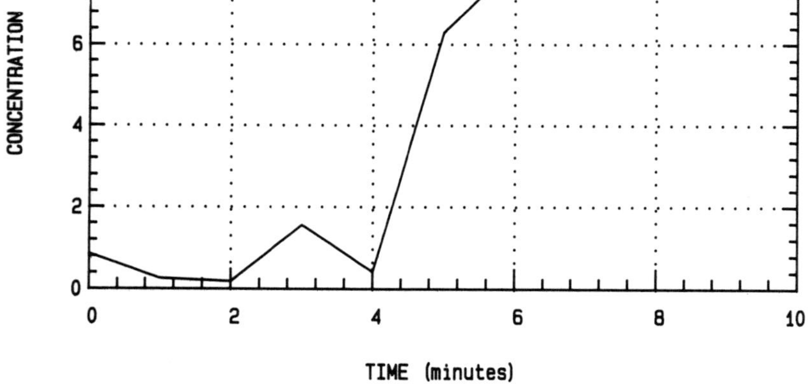

Fig. 13.8 — Phenol concentration curve based upon samples taken from the avoidance–preference chamber during a test with steelhead.

of 10 fish, while the bars define the standard deviation. Using steelhead, the elapsed time from the point of phenol injection to the statistically significant change in linear velocity was about seven minutes. The phenol level had reached about 8 mg/l in the avoidance–preference chamber. Speed increased rapidly when the phenol first entered the test chamber and then decreased below the level of the control (Fig. 13.9A). The vertical position of the fish also changed significantly (Fig. 13.9B) where

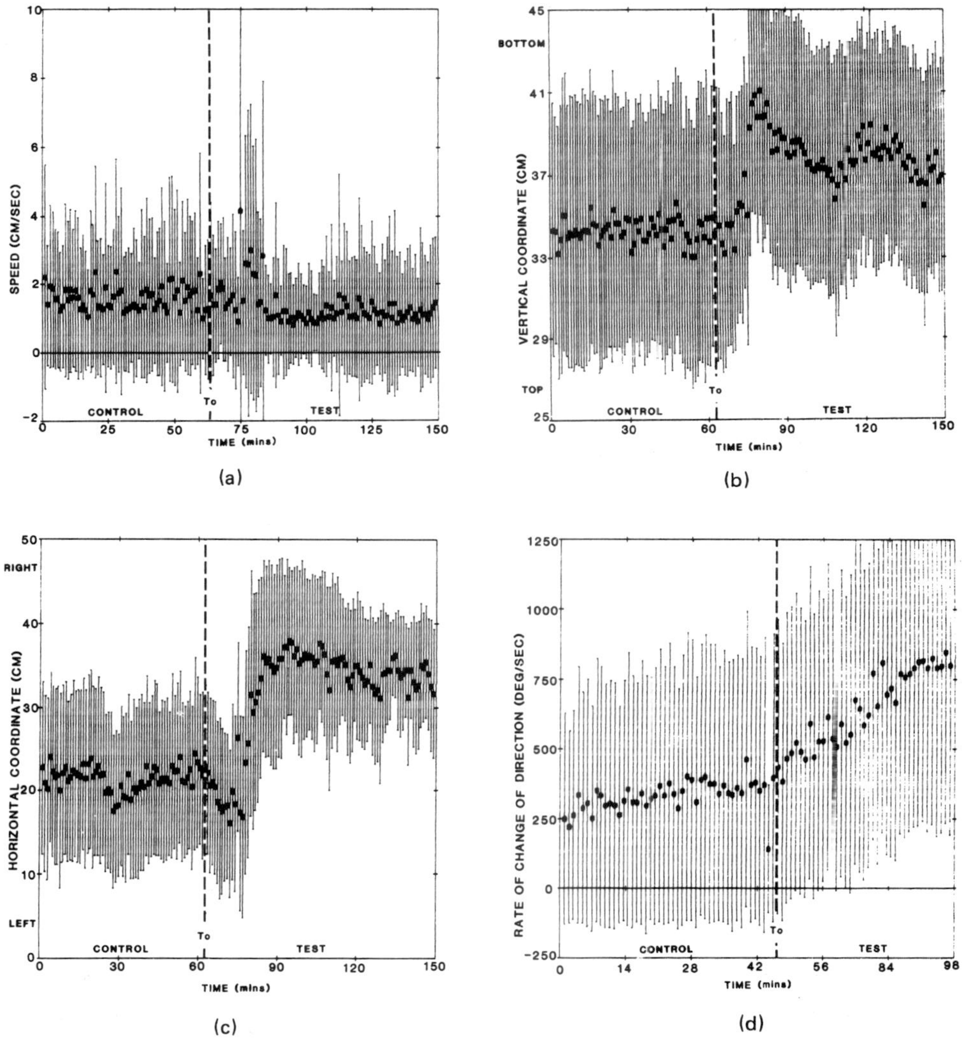

Fig. 13.9 — Plots of the change in the mean and standard deviation for various behavioral parameters measured during steelhead exposed to phenol: (a) linear velocity; (b) vertical coordinates; (c) horizontal coordinates; (d) rate of change of direction.

the fish moved away from the incoming phenol (phenol entered from the top of the chamber). Changes in the horizontal position followed that of the vertical position with the fish moving away from the side into which the phenol was injected (Fig. 13.9C). Statistically significant response levels were established for the data sets by a two sample analysis ($p \leq 0.05$, Statistical Graphics Corporation). The control response data were compared to the test data and significance established between groups. The rate of change in direction also indicates a behavioral response to the phenol (Fig. 13.9D). This measurement shows the change in swimming direction and increases in angular movement of the fish as they sense the presence of the phenol. Rainbow trout showed a more marked difference in swimming speed (Fig. 13.10)

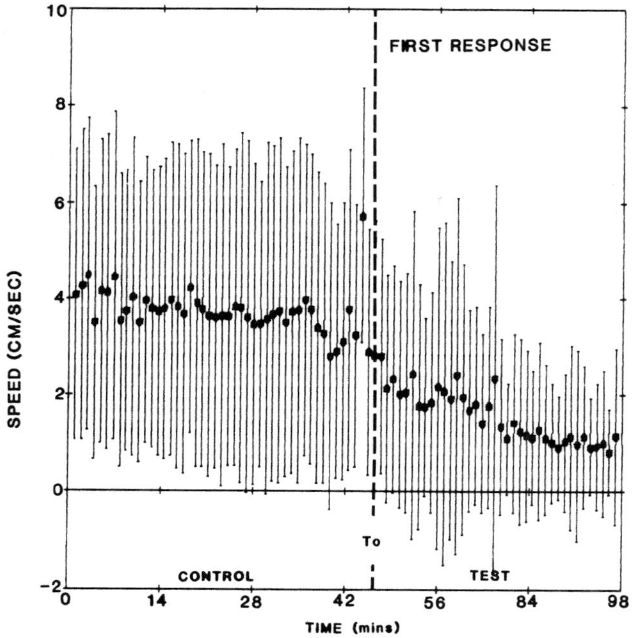

Fig. 13.10 — Plot of the change in the mean and standard deviation indicating a strong decrease in swimming speed for rainbow trout exposed to phenol.

when exposed to phenol. Another way of analyzing the response data is to plot a mean difference graph (Fig. 13.11). This analysis plots each data point in reference to the mean and graphically shows a change in horizontal position for steelhead exposed to phenol.

13.3.3 *Daphnia* response to phenol

Fig. 13.12 illustrates control behavioral variations over time. Each dot represents the mean of 20 organisms with the bars indicating standard deviations. Tests have been

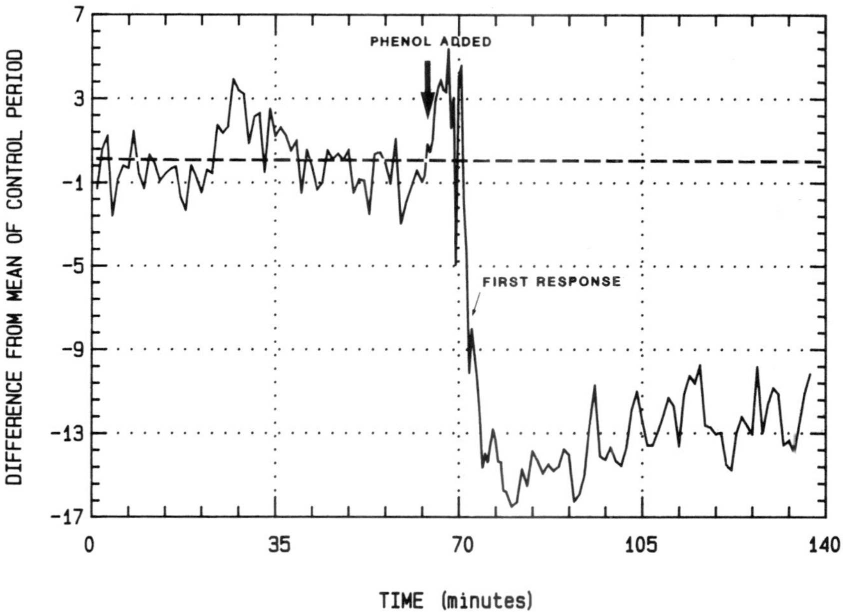

Fig. 13.11 — Plot of the mean difference from the control for steelhead exposed to phenol.

run for up to 164 hours with only two outliers. Linear velocity (Fig. 13.12) shows a tight data set with only one outlier significant at the 0.05 level.

Concentration curves were established for each exposure to phenol. Early tests used an LC_{50} value of 17 mg/l established for phenol in our laboratory. Later tests used about half that value or 8 mg/l (Fig. 13.13).

The set of *Daphnia* tests were conducted to determine response time and level of acclimation to multiple exposure of phenol using the same group of organisms. Fig. 13.14a shows the linear velocity increase of *Daphnia* when exposed to phenol. During the post-exposure (purging of phenol from the chamber) period swimming speed returned toward the pre-exposure level. Movement in the vertical plane changed (Fig. 13.14b) from a more random pattern in the pre-exposure period to a concentrated position on the right side of the chamber away from the phenol entry side (left). In both positional measurements (Fig. 13.14c), a return toward a more random movement occurred during the test solution purging period (post-exposure).

In all the *Daphnia* phenol tests the organisms responded to much lower concentrations of phenol than the fish tested. The lower level response was at 0.43 mg/l while the highest level reached was 2.0 mg/l (Table 13.1). Response times were also much shorter than the fish, averaging 1.8 min. The mean difference plot for linear velocity demonstrates the sharp increase in swimming speed in response to the phenol (Fig. 13.15). This same quick response can be seen in the mean difference for horizontal position where the *Daphnia* moved away from the entry point of the phenol.

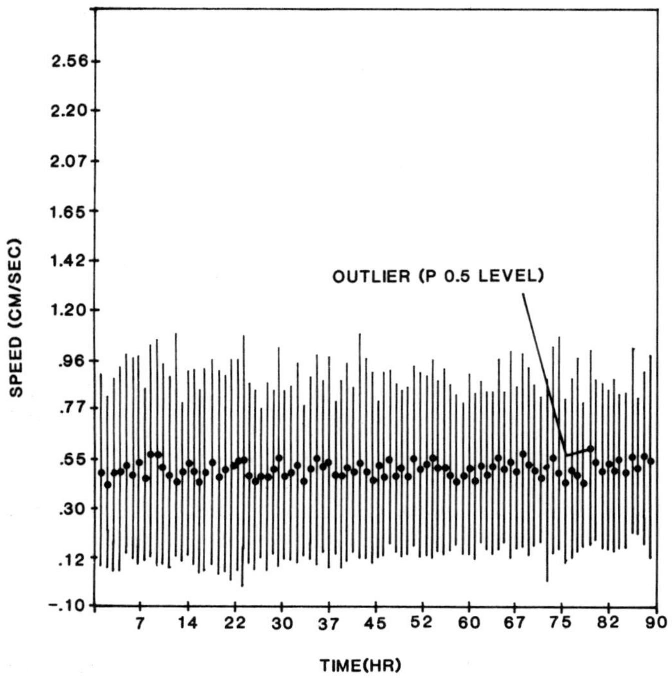

Fig. 13.12 — Plot of means and standard deviations for a long-term control test using *Daphnia* (speed).

When the pre-exposure and post-exposure test periods were compared, the test response was statistically significantly different. After 24 hours the post-exposure swimming speed had not returned to the same level as during the pre-exposure period; it remained slightly elevated. However, when the second dose of phenol was added, the level of response was statistically significantly different from the elevated post-exposure level. This indicates that, even though the organisms have been exposed to the phenol previously, when exposed again after 24 hours, they can respond strongly and still be detected.

13.4 DISCUSSION

Most of the early warning systems reviewed use fish gill-ventilatory response based upon bioelectric signals which are most likely generated by the muscles that move the buccal apparatus and the opercular plates of the fish (Cairns and Gruber, 1980). These systems use a single species and a single behavior-response characteristic. The system used in this paper is based upon a multiple-species approach that monitors actual behavior of the organisms. More recent commercially available systems have tried different approaches to early warning response (Aquatest, Grider System); however, they also use fish as the only test organism. We are testing a system using

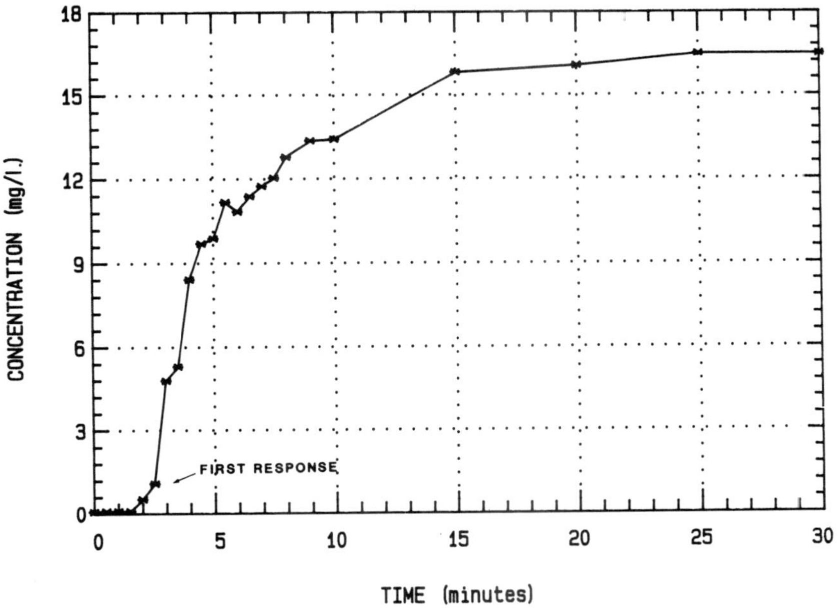

Fig. 13.13 — Phenol concentration curves based upon samples taken from the *Daphnia* chamber during testing.

both an invertebrate and a fish in the hopes that a more sensitive response to a wider range of potential toxicants can be developed. This system can also use a larger number of organisms (up to 50 individuals) which can be smaller (early developmental stage) in size than used in other systems.

In comparing the very preliminary fish-response data generated by this video system to existing early warning systems, the ventilatory response seems more sensitive. The Mk III Fish Monitor (Evans *et al.*, 1986) detected phenol at 2 mg/l with an alarm raised in 28 minutes. Using trout/steelhead the response time of seven minutes to 8 mg/l of phenol is faster but not as sensitive. The seven-minute response time is based upon the first statistically significant mean from the control where the Mk III time for alarm is based upon 10 consecutive excedences of the alarm threshold. Applying that same criteria to the video system, the response time would be 17 minutes. The Grieder System using a Nile River fish *Gnathonemus petersi* detected the presence of 0.2 mg/l cyanide after two hours of exposure and 0.5 mg/l copper after 5.5 hours of exposure. The fish does give off unintentional alarms during sleep periods (Ewen, 1987).

We are exploring more sensitive analytical methods to detect lower levels of response including following individual fish behavior rather than the mean of the entire group and/or applying different types of time-series analysis. However, although tracking individual fish during the response period may produce a more sensitive response to lower toxicant levels, care must be taken not to introduce

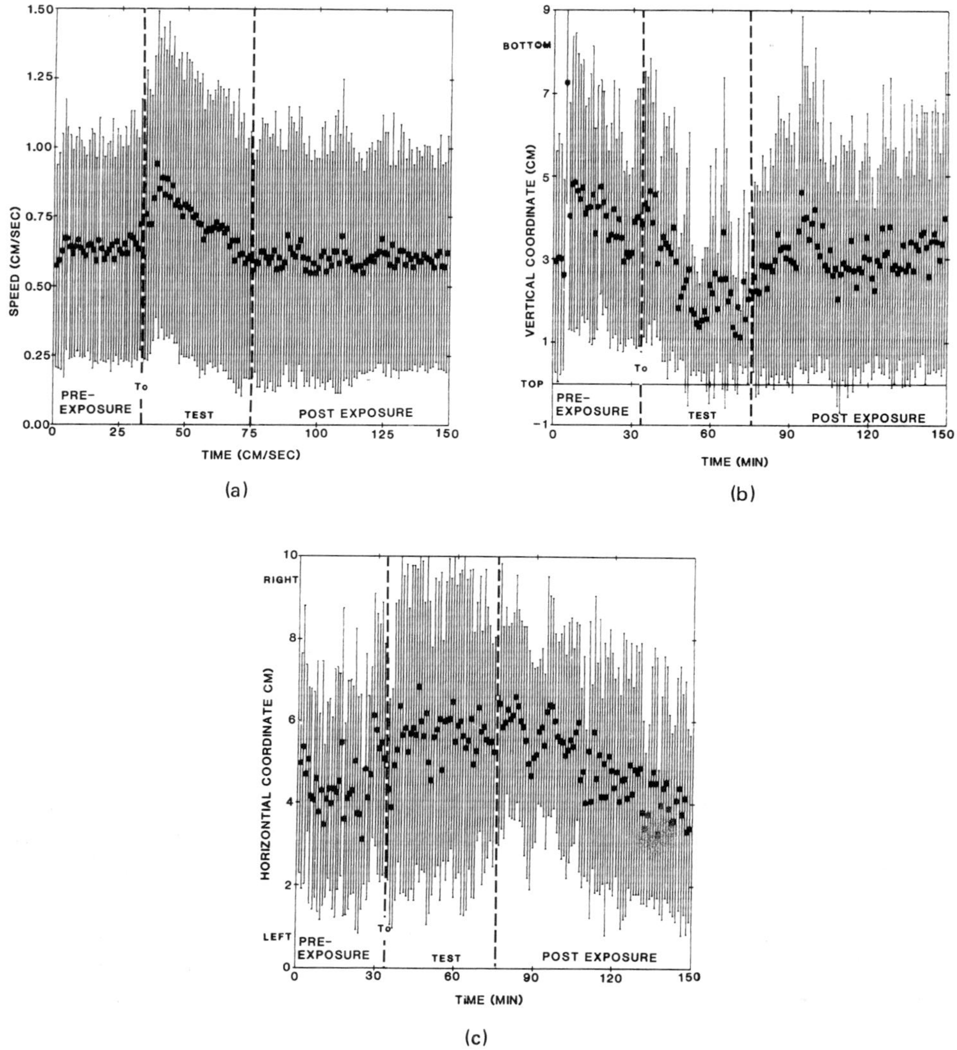

Fig. 13.14 — Plots of changes in various behavioral parameters for *Daphnia* exposed to phenol (results expressed as means and standard deviations): (a) speed; (b) vertical coordinates; (c) horizontal coordinates.

individual behavioral characteristics which could produce false alarms. Recently completed tests using fathead minnows *Pimephales promelas* show a response time of six minutes to 0.7 mg/l phenol by using different flow rates and chamber design.

Daphnia behavioral response to phenol as measured by our system showed fast response of 1.8 min for the first response and 6.8 min if 10 consecutive responses which are statistically significant from the control are used. The response time could be shortened by taking data more frequently. The behavioral response was induced by levels of phenol as low as 0.43 mg/l. Using a much earlier version of our system,

Table 13.1 — Behavioral response of *Daphnia* and salmonids to phenol

Test species	Response dilution (mg/l)	Vertical position	Horizontal position	Swimming speed	Detection time[a] (min)
Trout 1	10.8	(−)	(−)	(−)	11
Trout 2	10.6	(−)	(−)	(−)	13
Steelhead	8.0	(−)	(−)	(−)	7
Salmon	12.6	(−)	(−)	(+)	7.5
Daphnia 1	0.43	(−)	(−)	(+)	1.5
Daphnia 2	1.2	(−)	(−)	(+)	2.0[b]
Daphnia 3	0.8	(−)	(−)	(+)	1.9
Daphnia 4	1.6	(−)	(−)	(+)	2.1[b]
Daphnia 5	2.0	(−)	(−)	(+)	1.6

(−) Significantly decreased (speed); significant avoidance (position).
(+) Significantly increased (speed; significant attractant (position).
[a]Response time based upon significant change in speed.
[b]Second exposure of the same organisms (*Daphnia*) to phenol after 24 hours of purging with control water.

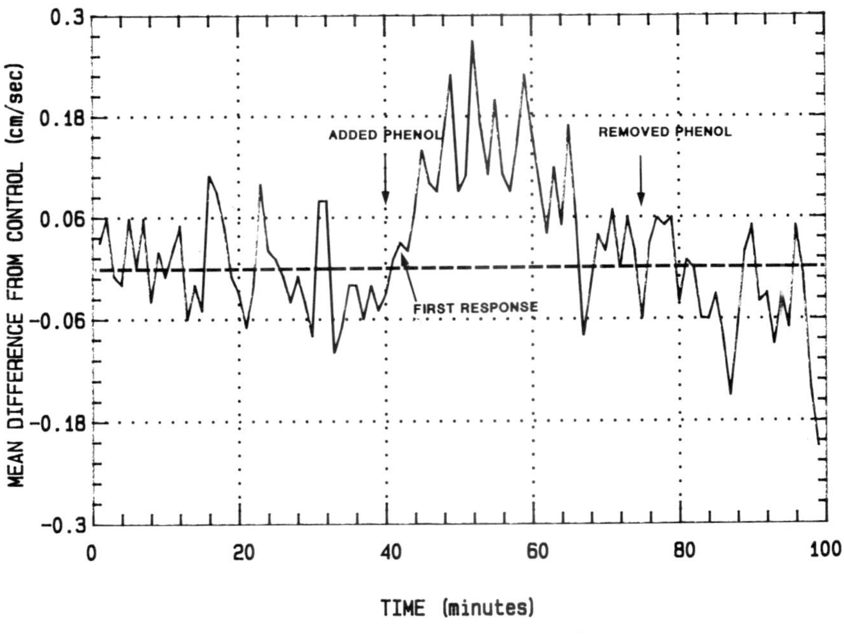

Fig. 13.15 — Plot of the mean difference from the control for *Daphnia* exposed to phenol (horizontal coordinates).

Miller *et al.* (1982) were able to detect copper at 18.5 ppb by the hyperactivity of barnacle nauplii.

The possible combination of continuous fish behavioral biomonitoring with the use of an invertebrate such as *Daphnia* can offer a potentially more sensitive system both in level and speed of detection and in sensitivity to a wider array of chemicals. Any number of test species could be used in the system, including both fresh-water and marine species. Our selection of test species was based upon availability, the size of the data base, previous use as a bioassay organism and the behavioral characteristics of the organisms. Brown (1976) noted the importance of the multispecies approach to ensure a detectable response to a wide selection of toxic compounds.

There are some operational problems with all early warning systems reviewed. Observed amplitudes in ventilatory response can be effected by changes in the dissolved salts and the direction of the test fish in relationship to the detection electrodes (Cairns *et al.*, 1981). The use of video images can be affected by seasonal variations in turbidity and contrast between the organism and the background. The turbidity problems can be greatly reduced by the use of filtration, although some regulatory agencies are concerned about filtration removing some toxicants. However, with early warning systems the quick response will most likely be to toxicants in solution. Materials sorbed to particulate matter may be unavailable for uptake by organism and thus not detected (van der Shalie, 1981). The field site used for this system has two types of in-line filters which have worked well during highly turbid flow periods, although some maintenance is required to keep the filters clean.

The importance of selecting organisms which have good contrast between the organism and the background is paramount with this system. Some invertebrates are almost transparent in water and thus are extremely difficult to digitize. Large, adult *Daphnia* fed with dark food can be digitized and juvenile fish illuminated from below are very easily followed by the system.

The organism must be active and in constant movement, as was noted by Sherer and Nowak (1973) in their work on measuring avoidance movements in fish. Early work done by the authors on juvenile channel catfish (*Ictalurus punctatus*) demonstrated that although catfish are sensitive to phenol, the control produced a difficult long-term pattern. The background noise caused by very infrequent activity followed by sharp, short locational changes and hyperactivity was extremely difficult to use in an early warning system.

Whereas many toxicological testing procedures require diel lighting regimens, the test procedures selected for these experiments were based upon continuous lighting. This reduces the sharp transitions between light and dark which produce increased locomotor activity and opercular rate (Drummond and Carlson, 1977).

Cairns and van der Schalie (1980) pointed out that a basic factor in the successful operation of continuous behavioral biomonitoring was the ability to detect deviations from normal conditions quickly and reliably. The Mk III Fish Monitor adopted four statistical methods: Cusum, Mann–Whitney U, non-parametric regression and percentiles. Based upon the tests conducted for that system (Evans *et al.*, 1986), the percentile test performed most satisfactorily on the data series. Cairns *et al.* (1981) used time-series analysis in his gill ventilatory system. The behavioral data generated by our video system can be analyzed in many different ways. We are still exploring the best methods for inclusion in an early warning system context. We have used the Kolmogarov–Smirnov two-sample test, Mann–Whitney U, Mann Whitney Wilcoxon test, percentiles and numerous time-series analysis. While we have not made a final selection, the concept developed for the Mk III system, or some modification, seems to be a good starting point. Confidence limits or acceptable percentiles can be set with the computer system checking control data consistently by gathering and determining if each new data point falls within these preset boundaries. The sensitivity could be set based upon the number of consecutive data points that fall outside the allowed confidence limits. The control data developed for *Daphnia* indicate that the number of single outliers are not frequent and that for tests lasting up to 164 hours we had no two outliers in succession. The data developed from the field site have not been analyzed for inclusion in this paper; however, they should also serve as a good indication of control variability over time.

13.5 CONCLUSIONS

Preliminary laboratory tests with phenol and other toxicants indicate that the approach of following behavioral patterns with a video-based computer system holds promise for biomonitoring. A number of organisms (up to 50) can be followed simultaneously in the same field, which greatly increases the statistical power of a system used for behavioral monitoring.

Daphnia sensitivity levels and response times to phenol were lower than for the salmonids. Although the response levels and time were lower for the gill ventilatory response than for the salmonid overall behavioral response, *Daphnia* still exhibited faster and lower level responses than the ventilatory measurements reviewed from the literature. However, recently completed data on fathead minnows using different flow rates and chamber design indicate the fish response levels and response time can approach those of *Daphnia*.

Seven-day control data for *Daphnia* indicate a replicable and stable response base with few outliers. Control data for fish need to be expanded; however, preliminary results indicate salmonids tend to be more irregular over time in their control behavioral patterns than *Daphnia*. Size, sexual state and number of fish used in the test are factors. Visual sensitivity to outside stimulus is also a factor.

Multiple exposures of the same organisms (*Daphnia*) to phenol result in detection of the toxicant by the organism and a statistically significant response. Pre- and post-exposure data were significantly different, even after 24 hours of exposure to control water. However, this fact did not result in a no-detection response.

The video system could encounter difficulties with high turbidity levels in the exposure chambers. However, this problem can be alleviated with on-line filters to remove particulates.

ACKNOWLEDGEMENTS

We wish to thank Robert Klamt of the California Regional Water Quality Control Board, Bob Morison of the Sonoma Water Agency and John Greaves of Motion Analysis Corporation for all their help on this project. Special thanks to Bridgette Hejtmanek for her assistance in conducting the experiments and to Carol Lee for her help in editing and preparing some of the figures.

REFERENCES

Anonymous (1986) *Aquatest: a warning and testing system to detect automatically acutely poisonous substances in water and wastewater.* Quantum Science Limited, England.

Barera, T. and Adams, W. J. (1983) Resolving some protocol questions about *Daphnia* acute toxicity tests. *Aquatic toxicity and hazard assessment, sixth symposium,* ASTM STP 802, pp. 509–518. American Society for Testing and Materials, Philadelphia.

Besch, W. K., Kemball, A., Meger-Warden, K., and Scharf, B. (1977) *A biological monitoring system employing rheotaxis of fish, in biological monitoring of water and effluent quality.* ASTM STP 607, pp. 56–74. American Society for Testing and Materials, Philadelphia.

Brown, V. M. (1976) Advances in testing the toxicity of substances to fish. *Cheng Ind.* **4** 143–149.

Buikema, A. L., Geiger, J. G., and Lee, D. R. (1980). *Daphnia* toxicity tests. In *Aquatic invertebrate bioassays.* ASTM STP 715, pp. 48–69. American Society for Testing and Materials, Philadelphia.

Cairns Jr, J. and Gruber D. (1980) A comparison of methods and instrumentation of biological early warning systems. *Wat. Resourc. Bull.* **16** (2).

Cairns Jr, J., Thompson, K. W., and Hendricks, A. C. (1981) *Effects of fluctuating, sublethal applications of heavy metal solutions upon the gill ventilatory response of bluegills* (Lepomis macrochirus). EPA 600/53-81-003. Environmental Protection Authority, Washington, DC.

Cairns Jr, J. and van der Schalie, W. H. (1980) Biological Monitoring Part I — Early Warning Systems. *Wat. Res.* **14** 1179–1196.

De Graeve, G. M., Geiger, D. L., Meyer, J. S., and Bergman, H. L. (1980) Acute and embryo-larval toxicity of phenolic compounds to aquatic biota. *Arch, Environ. Contam. Toxicol.* **9** 557–568.

Drummond, R. A. and Carlson, R. W. (1977) *Procedures for measuring cough (gill purge rates of fish).* EPA 600/3-77-133. Environmental Protection Authority, Washington, DC.

Evans, G. P., Johnson, D. and Withell, C. (1986) *Development of the WRC MK III Fish Monitor: description of the system and its response to some commonly encountered pollutants.* Publication Environ. TR233. Water Research Centre, Medmenham, Marlow.

Ewen, R. (1987) *Biological testing for toxicity control in open waters.* Endress and Hauser, Germany.

Giattina, J. D., Gorton, R. R. and Steven, D. G. (1982) The avoidance of copper and nickel by rainbow trout as monitored by a computer-based acquisition system. *Trans. Am. Fish Soc.* **111** 491–504.

Greaves, J. O. B. and Wilson, R. S. (1980) *Development of an interactive system to study sublethal effects of pollutants on the behavior of organisms.* US EPA 600/3-80-10, Washington, DC.

Gruber, D., Cairns Jr, J. and Hendricks, A. C. (1981) Computerized biological monitoring for demonstrating wastewater discharge. *Journ. Water Poll. Cont. Fed.* **53** 505–511.

Hasselrot, J. B. (1975) Bioassay methods of the national Swedish environmental protection board. *J. Wat. Pollut. Control Fed.* **47** 851–857.

Ishio, S. (1964) Behavior of fish exposed to toxic substances. In *Advances in Water Pollution Research,* Proceedings of the 2nd International Conference on Water Pollution Research, Tokyo, vol. 1, pp. 19–33, Pergamon Press, London.

Jackson, H. W. and Brungs, W. A. (1966) Biomonitoring industrial effluents. *Ind. Wat. Engng.* **45** 14–18.

Jones, J. R. E. (1951) The reaction of the minnow *Phoxinus phoxinus* (L) to solutions of phenol, ortho-cresol and para-cresol. *J. Exp. Biol.* **28** 261.

Klamt, R. (1985) *Workplan for Water Quality Management Planning Program (Section 205J) on toxic substance detection and early warning for the Russian River.* Water Quality Control Board, North Coast Region.

Klaverkamp, J. F., Kenney, A., Harrison, S. E., and Darrel, R. (1985) An evaluation of phenol and sodium azide as reference toxicants in rainbow trout. *Proceedings of the Second Annual Aquatic Toxicity Workshop.*

Larrick, S. R., Dickson, K. L., Cherry, D. S., and Cairns Jr, J. (1978) Determining fish avoidance of polluted water. *Hydrobiologia* **61** 257–265.

Lewis, M. A. (1981) Effects of loading density on the acute toxicities of surfactants, copper, and phenol to *Daphnia magna, Straus. Arch. Environ. Contam. Toxicol.* **12** 51–55.

Lubinski, K. S., Dickson, K. L., and Cairns Jr, J. (1977) Microprocessor-based interface converts video signal for object tracking. *Comput. Design.* Dec., 81–87.

Miller, D. C., Lang, W. H., Greaves, J. O. B., and Wilson, R. S. (1982) *Investigations in aquatic behavioral toxicology using a computerized video quantification system.* ASTM STP 766. American Society for Testing and Materials, Philadelphia.

Ochysnski, (1960) The absorbtimetric determination of phenol. *Analyst* **85** 278.

Sherer, E. and Nowak, S. (1973) Apparatus for recording avoidance movement of fish. *J. Fish Res. Board Con.* **30** 1594–1596.

van der Schalie, W. H. (1981) *Utilization of aquatic organisms for continuously monitoring the toxicity of industrial waste effluents.* Paper 22. U.S. Army Medical Bioengineering Research and Development Lab, Fort Detrick, Maryland.

Westlake, G. F. and Lubinski, K. S. (1976) A chamber to monitor the locomotor behavior of free swimming aquatic organisms exposed to simulated spill. *Proceedings of a National Conference on Control of Hazardous Materials Spill, New Orleans, LA.*

Index